KENNEDY SPACE CENTER
GATEWAY TO SPACE

We're Behind You, Discovery!

KENNEDY SPACE CENTER
GATEWAY TO SPACE

David West Reynolds

FIREFLY BOOKS

A Firefly Book

Published by Firefly Books Ltd. 2006

FIRST PRINTING

Publisher Cataloging-in-Publication Data (U.S.)
Reynolds, David West.
    Kennedy Space Center : gateway to space /
David West Reynolds.
[248] p. : col. photos. ; cm.
Includes: bibliographical references and index.
Summary: A detailed history of the Kennedy Space
Center, including early rocket and launch facilities
development, the missile race, the Apollo program, the
Shuttle, the Space Station and the Hubble Telescope.
ISBN-13: 978-1-55407-039-8
ISBN-10: 1-55407-039-2
1. John F. Kennedy Space Center – History. I. Title.
629.47/8/0975927 dc22   TL4027.F52R49 2006

Library and Archives Canada Cataloguing in Publication
Reynolds, David West
    Kennedy Space Center : gateway to space /
David West Reynolds.
Includes bibliographical references and index.
ISBN-13: 978-1-55407-039-8
ISBN-10: 1-55407-039-2
1. John F. Kennedy Space Center—History.
2. Launch complexes (Astronautics)—Florida.
3. Astronautics—United States—History. I. Title.

TL4027.F52J64 2006   629.47'80975924
C2006-902420-0

Published in the United States by
Firefly Books (U.S.) Inc.
P.O. Box 1338, Ellicott Station
Buffalo, New York 14205

Published in Canada by
Firefly Books Ltd.
66 Leek Crescent
Richmond Hill, Ontario L4B 1H1

Cover and interior design by Gareth Lind / LIND design

Printed in Belgium

The publisher gratefully acknowledges the financial
support for our publishing program by
the Government of Canada through the
Book Publishing Industry Development Program.

Atlantic Ocean

Merritt Island

NASA Airstrip

Pad 39B

Pad 39A

Kennedy Space
Center VAB

Indian River

Cape Canaveral
Air Force Station

*Banana
River*

Acknowledgments

T HE AUTHOR IS grateful to the many people who contributed their time and expertise to this book, from tour guides to technical specialists to the many public affairs officials who kindly arranged access to locations, equipment and personnel. Among those who helped I would like to specially recognize and thank the following individuals for their valued assistance:

Historical interviews
Helmut Zoike and Erika Zoike
(A4 launch sequence)
Konrad Dannenberg and Jackie
Dannenberg (von Braun team history)
Norris C. Gray, Civilian Safety
Officer (Bumper 8 and early
launches at the Cape)
Orion Reed (Navaho program)
Steve Bullock (VAB), John Neilon (ELVs)

KSC Media Office
Manny R. Virata, Jr. (KSC arrangements)
Margaret Persinger (images)
Kay Grinter (research)
Bill Johnson (LC-39 inspection)
Elaine Liston, Archivist (older images)
Bruce Buckingham

KSC site inspections
Thurston H. Vickery, Manager,
Transporter Group, United
Space Alliance
Kelvin M. Manning, OV-104 Atlantis
Vehicle Manager,
NASA (LCC and OPF)
Todd M. Konieczki, Manager GSS
Crane Operations, United
Space Alliance (VAB)

**Phaeton Group KSC Site
Inspection Field Team**
Hugh Williams, Mechanical Engineer
Cara Evangelista, Information
Officer and Media Assistant

White Sands Missile Range
George M. House, Curator, New
Mexico Museum
of Space History
Jim Eckles, PAO White
Sands Missile Range

Vandenberg Air Force Base
Robert S. Villanueva, Boeing
Brian Hill, Sr. Airman and PAO
Jim Benson, CEO SpaceDev

Scaled Composites/SpaceShipOne
Burt Rutan, Mike Melvill, Stu Williams

Pegasus XL
Barron Beneski, Orbital Sciences
Corp. Public Relations

Contact Assistance
Irene Willhite, Curator, U.S. Space
& Rocket Center, Huntsville
Ross B. Tierney

General Research
Al Hartmann
Capt. Matthew Cobb, Magnolia
Firehouse, Larkspur, Calif.

Special thanks also to Roger Launius at the Smithsonian National Air & Space Museum for inspiration and collegiality, Glasgow space authority Dave Harland for superb technical review and kind support, and Chuck Hyman for bringing me aboard this project.

To all the team members whose hard work, high standards
and determination have made the Cape rocketlands
a proving ground not just of hardware, but of the human spirit.

Contents

Foreword 14

Prologue 17

Section One: Gateway to Space

1 **Rocketland** 21
2 **Countdown** 39

Section Two: Road to the Cape

3 **Prototype** 45
4 **Location** 57

Section Three: Race for the Heavens

5 **The Challenge** 69
6 **Mercury** 79
7 **Gemini** 97

Section Four: Journey to the Moon

8 **Saturn** 115
9 **Moonshot** 139
10 **Skylab and Soyuz** 153

Section Five: The Shuttle

11 **Vision Quest** 167
12 **The Space Shuttle** 177

Section Six: Beyond the Shuttle

13 **Probes and Satellites** 205
14 **Gateway to the Future** 217

KSC Timeline 230

Recommended Reading 237

Index 241

Foreword

THE CAPE WAS a military installation that tested rockets intended to help us win the Cold War. The newcomers (as we were called when we first arrived) brought along a civilian space program intended to put a human (called an astronaut) into space.

The oldtimers (as we called the military types) at the Cape were assembling and fueling rockets, putting a "nose cone" on top and launching the rocket in the hope of delivering an explosive warhead to a faraway target. Each rocket was a one-shot deal.

The newcomers brought with us a corrugated nose cone, which we called a "spacecraft," to be put on top of a rocket. We proposed to put a chimpanzee and later a human in the nose cone, and hoped to bring that cargo back alive. The oldtimers were amazed. Did we not see the rocket failures and blowups week after week? We must be crazy! The oldtimers shook their heads and wished us luck.

There were even seven volunteers (astronauts) who were willing to ride in these funny nose cones. At each of the meetings with the Cape people we tried to convince them that our nose cone was a spacecraft. After each rocket blowup we tried to find out what went wrong, but were told that we did not have a "need to know" – and that the test results were classified top secret.

Despite the setbacks, we believed that we could accomplish our project to put a man in space. We carefully assembled, served and launched several unmanned test rockets and recovered the spacecraft. We made mistakes, but learned from them and improved the spacecraft for the next launch.

We launched two chimpanzees and brought them back alive, a feat that increased our approval rating with the oldtimers. Newspaper writers even dropped the "crazy scientists" label they had given us.

The launches of Sputnik and the first Russian cosmonaut took some wind out of our sails, but with the launch of Shepard and the rest of the Mercury astronauts, we began to take off.

The Gemini program let us move ahead of the Russians in spacecraft capabilities and operational experience. With new facilities built for the civilian space programs we entered the Apollo era, which culminated in landing a man on the moon and returning him safely to Earth.

The newcomers of the early Cape days are now the oldtimers. We look forward to the next generation advancing our space explorations.

— Guenter F. Wendt

Guenter F. Wendt, the last man seen by the flight crews prior to liftoff, was responsible for spacecraft launch preparations for all Mercury, Gemini and manned Apollo and Skylab flights.

Prologue

THE MISSION OBJECTIVES are decided elsewhere. The spacecraft are designed and built elsewhere. Today the astronauts are trained elsewhere, and their mission is controlled elsewhere. But the gateway to space lies here. All the elements of America's space program come together at Cape Canaveral's Kennedy Space Center.

After everyone else in the vast NASA spaceflight organization has done his or her job, the Kennedy Space Center team bears the tremendous responsibility of being the last ones to make sure that all the work has been done right and that every component of a spacecraft will function properly. There will be no one after them to check their work or catch any mistakes.

Space rockets require extremely elaborate preparations. Each rocket type must have a specialized launch pad tailored to its precise needs. Without painstaking preparatory operations and supporting equipment to match, a rocket is useless. The most sophisticated rockets in the world require the most sophisticated launch facilities, and the finest are found at the legendary places known as Cape Canaveral and Kennedy Space Center.

← As the swing arms move away, a plume of flame signals the liftoff of Apollo 11 carrying astronauts Neil Armstrong, Michael Collins and Edwin Aldrin Jr. to the moon. July 16, 1969.

GATEWAY TO SPACE

The most sophisticated rockets in the world require the most sophisticated launch facilities, and the finest are found at the legendary places known as Cape Canaveral and Kennedy Space Center.

ROCKETLAND

CAPE CANAVERAL IS a point of land that juts out from Florida's east coast into the Atlantic Ocean so prominently that you can easily spot it from orbit. The Cape is one of the few geographic features that show up on the earliest Spanish maps of Florida. It served as a landmark for the treasure galleons making their way up through the Bahamas Channel. Upon sighting the Cape, ships would steer to starboard, and head east into the open ocean, toward the continents beyond the horizon. Many of these ships ran afoul of the shoals lurking in these waters, and their remains litter the bottom. Ancient timbers and artifacts mingle on the seafloor with crashed missiles and rocket parts from launches gone awry over the last half-century. The Cape remains a threshold of risk, as is any place where challengers dare to step away from one world into another.

← The Space Shuttle *Columbia* on a mission to the Hubble Space Telescope photographed over the Rocket Garden at KSC Visitors Complex. 2002.

KENNEDY SPACE CENTER

NASA's KENNEDY SPACE Center (KSC) sprawls across almost 140,000 acres on the northern portion of the Cape. At the center of this complex stands one of the largest and most distinctive buildings in the world: the Vehicle Assembly Building (VAB), where titanic craft are assembled in gigantic interior bays and readied for space. Ranged around the VAB lie the landing strip for the space shuttle, the shuttle orbiter hangars, and the Launch Control Center. Away from the VAB a short superhighway runs for about three miles to the coast. This road was built for one special type of vehicle, the powerful crawler-transporters that haul NASA's mightiest spacecraft. The road splits and then ends at twin launch Pads 39A and 39B. Beyond here, the road leads straight up — into the air. Since Apollo 8 in December 1968, these two pads have launched every U.S. astronaut mission.

Strictly speaking, Kennedy Space Center is cut off from Cape Canaveral by rivers, and so lies technically on Merritt Island. KSC is nonetheless part of the greater rocketland many simply call "the Cape." NASA built its early pads to the south on Cape Canaveral proper, within the territory of the Cape Canaveral Air Force Station, but by the mid-1960s they were running out of room. The Apollo Saturn V was so large that NASA expanded north of the true Cape and bought much of Merritt Island for the moon rocket mega-pads. This Saturn V complex built for Apollo became the Kennedy Space Center of today.

Six days after President John F. Kennedy was assassinated in November 1963, his successor, Lyndon Johnson, ordered the renaming of Cape Canaveral and the NASA launch center there in honor of his fallen predecessor. Thus, the great moon landing missions were all launched from "Cape Kennedy." Florida public sentiment for the historic original name of the Cape led to its restoration in 1973, but the name of Kennedy Space Center continues to serve as a memorial to John F. Kennedy's pivotal support of the space program.

The Space Center bustles with the activity of thousands of people, from the Center director through battalions of engineers and technicians to legions of support workers: a team large enough to populate a small city. Working together, this group turns spacecraft components into rockets ready for launch, checks equipment in countless different ways, transports hazardous fuels, and carries out myriad other tasks to get these ships into the air on schedule. An extensive industrial park provides KSC with services ranging from payload preparation to parachute packing. Space Station components from around the world arrive for final checkout and preparation at the Space Station Processing Facility. The

← Launch Complex 39 is dominated by the Vehicle Assembly Building and the crawlerway leading to pads 39B (on the left) and 39A (on the right). 1998.

↑ Aerial view of the Vehicle Assembly Building. 1999.

→ Aerial view of Launch Pad 39B. 2005.

KENNEDY SPACE CENTER

Manned Spacecraft Operations Building contains astronaut quarters, dining rooms, and the prep area where crews suit up for their missions.

CAPE CANAVERAL AIR STATION

BORDERING KENNEDY SPACE Center on the south is Cape Canaveral Air Station, under the authority of the air force. This is a military zone distinct from NASA's civilian jurisdiction. The Cape is packed with 46 launch sites, dating all the way back to the first blastoff in 1950 at Launch Complex 3, the first pad completed. Until 1968, all astronauts were launched from the Cape since they originally rode on modified military missiles. Alan Shepard became America's first man in space after launching aboard a modified Redstone missile from Pad 5, and John Glenn became America's first man in orbit after taking off atop an adapted Atlas missile from Pad 14. Most of the Cape pads were built to launch missiles not space rockets, and the long string of complexes along the coast includes the ominous "missile row," where Atlas and Titan nuclear

These Southern Bald Eagles inhabit an enormous nest on Kennedy Parkway North, seemingly uninterested in the noise and activity around them.

intercontinental ballistic missiles (ICBMs) were tested during the Cold War.

Military missiles accounted for a great deal of activity at the Cape, especially during the 1960s. It was here that the Army launched the Redstone, America's first tactical ballistic missile, and the Pershing, the Redstone's solid rocket battlefield successor. The Navy also test fired more than 168 Polaris, Poseidon, and Trident missiles from pads here. Navy teams also loaded many more such missiles into submarines at Port Canaveral harbor and fired them from below the waves. The air force used the Cape to test and launch a series of ICBMs, from Atlas, Titan I and Titan II to today's Minuteman. Most of the historic pads are now deactivated and dismantled, but scattered among the old sites are the active pads that still see the fire and thunder of unmanned rocket liftoffs.

Today most launches are conducted for the purpose of sending up satellites. Beginning in 1989, the Global Positioning System constellation has gone up on Boeing Delta rockets from Launch Complex 17, and a steady stream of communications and reconnaissance satellites are launched on rockets such as the modern Lockheed Martin Atlas V at

Launch Complex 41.

Nothing goes up without a weather check and an approval from the Range Safety office indicating that conditions are stable and suitable. About 50 Super Loki weather sounding rockets per year speed upward from Launch Complex 47 to altitudes of almost 57 miles. They serve as advance scouts that measure wind and temperature data through the air column to confirm safe conditions for the launch of larger machines.

NASA and the armed forces have shared and traded use and ownership of several Cape pad complexes over the years. Today the air force oversees the entire Cape, so when NASA launches an unmanned rocket, it has to borrow a pad from them. Many of the personnel working such a launch, however, will draw their paychecks from neither NASA nor the air force; they work for aerospace corporations such as Boeing Lockheed under contract to government agencies or commercial customers.

NATIONAL WILDLIFE REFUGE

PALMETTO SCRUB, SHALLOW meandering rivers and marshland were all that the early Spanish explorers saw here, and this tranquil tropical terrain still makes up much of the Cape. Manatees drift slowly through brackish rivers and creeks, while alligators sun on sandbanks. Armadillos trot across clearings amid pine and scrub oak forests, and soaring bald eagles rule over the rich diversity of bird species living out among the launch pads or stopping through on migration routes. Rangers keep watch on over 500 species of wildlife. The U.S. Park Service, which administers the Merritt Island National Wildlife Refuge established in 1963, guards 140,000 acres of this largely wetland environment. The Refuge serves a dual purpose: it protects the ecology of the wetland environment, and it provides a valuable safety and security buffer zone around the rocket sites. Between the occasional thunder blasts of rocket launches, this land is peaceful and serene, lush and abundant, and its animal life includes 15 endangered and threatened species.

OUR MINUTE SECOND

COUNTDOWN

2

THE SHEER MAGNITUDE of the facilities of Kennedy Space Center generates a feeling of awe in most visitors. So much equipment and so many facilities are involved at any launch pad that the sites are better known as "launch complexes." With their widely different configurations and sizes, they can seem far beyond the comprehension of the average person. The fact is, however, that even the gigantic Apollo/Space Shuttle facilities are really just extra-large versions of the same basic facilities needed by virtually any rocket, even the model rockets you can buy at a hobby store. Once you know what to look for, you can begin to appreciate the ways in which engineers have solved consistent problems in different ways over the years with ever-evolving rocket systems. This section introduces the basic elements of the launch complexes, the components of which form the vocabulary of the rest of this book.

← A digital clock at the press site counts the seconds since ignition of the Saturn rocket that lifts Apollo 13 into space. 1970.

The Aero Spacelines B377PG *Pregnant Guppy* sits on the ramp at Dryden Air Force Base awaiting flight tests and pilot evaluation. 1962.

ARRIVAL BY LAND – ROAD AND RAIL

Since rockets are not manufactured at the Cape, most space vehicles arrive as separate components that must be assembled and integrated near their launch site. The earliest V-2 rockets were trucked in on army tractor-trailers. When railway lines were brought in to serve the developing Cape, trains began to carry equipment as well. Railways still form a vital link in the space shuttle program, bringing in the craft's white "outrigger rocket" booster segments from their manufacturer, Thiokol, far across the country in Utah.

ARRIVAL BY AIR – THE SKID STRIP AND THE GUPPIES

Many rocket components have been flown to the airstrip known as the Skid Strip, located in the center of the Cape. Military cargo planes unloaded cargoes such as the first Gemini-Titan II rocket stages here. Shipping rockets from their manufacturers to the Cape presented a greater and greater challenge as components got larger over time. When even the smallest Apollo Saturn rocket stages got too big to fit in any standard cargo planes, NASA was looking at having to wait for these stages to be

At the Shuttle Landing Facility,
sections of the International Space
Station fit comfortably inside the
Super Guppy's 24-foot diameter
fuselage. 2002.

shipped by water to the Cape all the way from manufacturer Douglas Aircraft's factory in California. The long trip through Panama would take 18 to 25 days and threatened to wreck early launch schedules when testing problems delayed shipment.

But American entrepreneur Jack Conroy decided he could save NASA precious time in shipping Saturn upper stages by adding a gigantic balloon-like cargo hold to the back of an existing plane. The entire nose section of the plane hinged open at one side to give access to the cargo bay. Conroy founded Aero Spacelines and backed his idea with his own money. The outlandish result was one of the largest airplanes in the world. It looked ludicrously unflyable and won the unflattering nickname *Pregnant Guppy*. It drained Conroy's assets and he could not afford to finish its interior. Out of cash and desperate, Conroy flew the plane to the NASA headquarters of Apollo rocket mastermind Wernher von Braun in Huntsville, Alabama. Detractors and skeptics scoffed at the ungainly aircraft, but von Braun liked the boldness of Conroy's vision enough to climb aboard the bizarre creation for a demonstration flight. By the time Conroy's plane returned to the ground, the delighted von Braun was sold, and the *Pregnant Guppy* entered service for NASA in 1963. This extraordinary aircraft quickly put NASA schedules back on track when it flew the essential upper stage of the first Saturn I rocket from California to Cape Canaveral in just 18 hours. Over the next seventeen years, the *Pregnant Guppy* and its even larger sister ship the *Super Guppy* would go on to save NASA many months in delivery time, making a vital contribution to meeting the Apollo deadline for the moon.

In addition to Saturn upper stages, the Guppies flew many other major payloads, including Gemini rocket stages from the Martin Company in Maryland, Saturn rocket computer and instrument "brain" unit rings, and components of the Skylab space station: the core workshop, the Apollo Telescope Mount that gave Skylab its eyes into space and the multiple docking module. Today's descendant of the pioneering Guppy planes is NASA's *Super Guppy*, a European-built version of the American originals, which now carries International Space Station components. The *Super Guppy* still lands at the Skid Strip.

A barge carries the huge second stage of the Saturn rocket to the foot of the VAB. 1968.

ARRIVAL BY SEA – THE BARGE CANALS

The giant Saturn V Moon rocket lower stages were 33 feet in diameter, so large that not even the *Super Guppy* could carry them. These had to be brought in by barge. The Saturn V second stages, manufactured like the S-IVs by Douglas in Huntington Beach, California, were shipped down the coast of Mexico, through the Panama Canal, and up the Mississippi. At the Mississippi Test Facility (now renamed the Stennis Space Center), they were strapped down and tested at giant firing stands located out in the swampy wilderness. When approved, the stages were barged back down the river, back out into the Gulf of Mexico, through the Keys and up the east coast of Florida to Port Canaveral. From here a canal route led straight to the heart of Kennedy Space Center, almost to the very foot of the Vehicle Assembly Building (VAB), to a turnaround basin where the barges off-loaded their colossal cargoes.

Some of the smaller unmanned rockets make similar sea journeys today. Boeing's new Delta IV rockets arrive at the Cape in a specially designed, custom-made cargo ship, the *Delta Mariner*, which brings them in pieces down the Tennessee River from their construction site in Decatur, Alabama. The 312-foot *Delta Mariner*, commissioned in 2000, is built as a hybrid riverboat/oceangoing ship, offering 15 knots on the open sea, but drawing a mere nine feet for maneuvering in shallow canals and rivers.

ASSEMBLY BUILDINGS

Once the rocket components arrive at the launch area, they need to be assembled and scrutinized. The Cape's early assembly buildings were the "alphabet hangars," as each was assigned a letter for its name. These hangars, most constructed to a standardized design, were built one after another in what became a sizable industrial park at the Cape. The general-purpose buildings provided test and checkout facilities, cranes, pressurized gases and any other needed equipment. Hangar S served famously during the Mercury program for both space capsule checkout and sleeping barracks for those who would sit in space capsule seats – both apes and astronauts. As rockets got more complicated and their needs more elaborate, the hangars became filled with very specific equipment for each rocket.

During the Apollo program, the rockets got so large it was no longer possible to fit them in the old alphabet hangars. They had to assemble the Saturn IB right on the pad, which meant exposing the rocket to corrosive atmosphere for months while it was put together and inspected. For the ultimate rocket, the Saturn V, a special assembly building was created: Kennedy Space Center's towering VAB, the ultimate rocket garage. In this colossal building, 250-ton cranes stacked moon rocket stages one atop the other and linked them together. Nowadays crane operators pluck space

Space Shuttle *Discovery* mated to the ET and sitting on the Mobile Launch Platform begins to leave the VAB on its journey to Launch Pad 39B. 2005.

Inside the VAB, the nose cone for the Lunar Module is being placed into position. 1967.

shuttle orbiters off the ground and carefully rotate them into vertical position for joining to their booster rockets. The VAB is the greatest land monument of the Apollo program and continues to serve as an integral part of Kennedy Space Center operations today.

The air force built its smaller version of the VAB, the Vertical Integration Building (or VIB), at its Titan III launch complex at 40/41.

TRANSPORT

The earliest rockets to be launched at the Cape, the V-2s, were carried to their site by special trucks that used hydraulics to lift the rockets up to vertical and then set them down on the pads. Gemini rockets were brought in stages on trailers to their pads and assembled there. Saturn IBs were also brought in by trailer. When it came time for the Saturn V, the individual stages on their own could just barely be managed by large trailers, and the entire assembled rocket was so unbelievably huge that twin crawler-transporters were built, along with a special road called the "crawlerway," to move the rocket from the VAB to the launch pad. Smaller rockets could get around on rails, as they did at the 40/41 Titan III facility, which delivered assembled rockets from the assembly building to the pads on a continuous rail system.

LAUNCH PAD

Every rocket needs a launch pad built specifically for its needs. That's why there are so many different pads at the Cape. Each pad has to be equipped with special fittings to fuel and prepare the rocket, to support it properly before launch, and to deflect its launch blast. The pads take many different forms, but they all serve similar purposes.

A rocket needs to sit firmly on a launch stand. Usually this requires a few fixtures around the bottom of the rocket, where it is bolted to a pedestal over the pad, allowing the engine bells to hang freely. The bolts are cut at the moment of liftoff, but until then they hold the rocket firmly in place as the engines are run up to full thrust. The bolts also prevent high winds from blowing the rocket over during preparation.

Power, propellants and monitor connections must be run out to the pad in channels, often placed underground for protection. Cryogenic liquefied gases at hundreds of degrees below zero will shrink and crack any ordinary plumbing, so special tubing is required for transporting exotic fluids like liquid oxygen and liquid hydrogen. Cape engineers commonly use double-jacketed stainless steel tubing. The space between the tube walls is sucked almost clean of air to leave near-vacuum pressure – 1/76 of atmospheric pressure – providing excellent insulation.

Fuel and the oxidizer that mixes with it are both known as propellants. Supplies of both are kept on-site in big tanks; they are generally located at the very edge of the pad area to minimize the danger of explosions.

Before the propellants are loaded, the systems are purged with a dry, inert gas, such as compressed nitrogen, to clear out anything that may be in the pipes or tankage. Storage tanks of purging gas are tucked into one side of the Apollo pads, which serve the shuttle today.

UMBILICAL TOWER

A rocket is placed on its pad empty, and it needs to be fueled before liftoff. The umbilical tower is a scaffolding to support the various hoses that need to be hooked into the rocket to load the propellants. As well as fuel, other fluids and compressed gases may flow into the rocket to pressurize its tanks and charge up its gyroscopes and other systems. Electrical cables provide power and checkout hookups for engineering monitors.

The umbilical tower connections must be maintained until launch, but at that point they must also flawlessly disconnect. For the V-2 this was no big deal as the cables were simply yanked off by the movement of the rocket. For the giant Saturn V moon rocket, umbilical arms so large you could drive a car through them had to be disengaged with powerful charges and then rapidly swung out of the way as the moon rocket rose off the pad. The umbilical tower gets exposed to the force of the launch

The flame trench on Launch Pad 39B is 490 feet long and 40 feet high. Space Shuttle *Columbia* is on the pad. 1999.

blast, and acoustic shock from the raw noise can shake the entire structure like an earthquake. A built-in firehose drenching system cuts down the damage of a blastoff but, even so, a tower typically has to be refurbished after every launch so violent is the event. Connections, hoses and pipes all have to be checked and replaced or cleaned up.

GANTRY

Once the rocket is set up on its stand, it needs to be checked exhaustively. Technicians need access to nearly every part of the rocket to run their tests, perform maintenance and make repairs. A structure in addition to the umbilical tower is usually necessary. The service structure or gantry, as Cape workers call it, provides this access. The Cape's first gantry was a painter's scaffold for the launch of Bumper 8. The biggest gantry was the service tower built for the Saturn V, which was so large that it had to be carried into place by the same massive crawler that carried the moon rocket out to the pad. When the technicians have finished and the rocket is ready, the service tower is withdrawn from the way of the launch blast to protect it from damage.

↑ Firing Room #2 in the Launch Control Center during tests for Apollo 12. 1969.

→ Houston Mission Control directs the Space Shuttle *Discovery* as it docks with the International Space Station. 2005.

WHITE ROOM

During installation and checkout, the spacecraft at the top of a rocket needs special protection from the wet Cape weather. Without climate control, millions of dollars' worth of equipment can be ruined from exposure. A white room provides this protection, wrapping the spacecraft in a special enclosure at the top of the gantry.

The first white room was jerry-rigged during the early phases of Project Mercury. Two pad crewmen took the initiative to protect the Mercury spacecraft from water damage, stringing up plastic sheeting on the open gantry tower. Director Kurt Debus' operations team got the message and soon had the upper sections of the gantry professionally enclosed and skinned with translucent panels. This was the origin of the White Rooms that have become a standard feature of rocket gantries. All astronauts have begun their journeys by boarding their vehicles from a White Room that serves as the final threshold between Earth and space.

FLAME TRENCH

When a rocket's engines ignite, the flame and blast must be deflected out of the rocket's way. If a rocket were to be launched over a flat pad, the blast could rebound right back into the engine and destroy the rocket. Consequently, underneath every large rocket lies a deflector of some kind to direct the flame out to the sides. The largest rockets have massive flame deflectors built into the pad that channel the blast through a special trench lined with fireproof brick. The flame deflector gets exposed to the full power of a launch blast, so it has to be built unbelievably strong. The deflector used for the Saturn V moon rocket was 42-feet tall, and even though it was coated with volcanic concrete, the rocket blasts still tore off three-quarters of an inch at every launch.

LAUNCH CONTROL CENTER

Every launch pad needs a control center. Over the years control centers have taken a wide variety of forms, beginning with a tarpaper shack for Bumper 8. For the first launch of a Lark winged missile in 1950, the team hid inside a spare army tank. The crews launching the Matador winged missile made do with a sandbagged foxhole. The sturdy pillbox-bunker blockhouse of the Redstone pads was succeeded by a long line of igloo-like domes that protected two floors of communications and control consoles, and instruments for the Atlas ICBMs and early Saturn rockets. The ultimate Launch Control Center (LCC) is the low black-and-white building alongside the VAB, which was created for Apollo and is still used today for the space shuttle. This is the finest and most inspiring incarnation of the concept that has ever been built.

Members of the Final Inspection Team on Launch Pad 39B return after testing the External Tank in preparation for a shuttle launch. 2005.

The Redstone pillbox sat so close to the rocket that the building required thick armored slot windows to allow for direct observation of the pad. Launch control chiefs in the Atlas and Saturn igloo blockhouses used real periscopes like submarine commanders to watch their rockets lift off. Later electronic developments allowed the launch control center to be sited a safer distance away from the action, so the Apollo LCC stood safely over three miles from the pad. It would survive even the worst disaster that a Saturn V could offer, so the designers were free to make it a statement of beauty as well as utility; instead of an ignominious igloo, NASA's current LCC looks "like the future is supposed to look."

The control center is the final element of a launch complex. When it has done its job, the rocket is on its way.

MISSION CONTROL

Monitoring and controlling the rocket while it is in flight is an entirely separate job from launching it and generally requires its own building full of people and computers. The information feeding the Mission Control monitors is relayed by the tracking range network of radar and communications stations that the rocket flies over after it leaves the pad.

In the days of Project Mercury, Mission Control was at the Cape along with the Launch Control Center. With increasing computer sophistication, Mission Control could be set up anywhere and linked to the rocket by a communications network. After the first Gemini mission, Mission Control was moved to Houston, where it has remained for all U.S. astronaut missions ever since. Other rockets, like the unmanned Atlas, still have their Mission Control rooms on-site at the Cape.

Finally, all these components of a space launch complex would be idle except for the unseen web of power, communications, command, instrumentation and control lines that weave the organs of the Cape into a functioning body. The U.S. Army Corps of Engineers, along with contractors, military personnel and NASA engineers, has been there from the beginning doing the hard groundwork that makes possible our reach for the stars.

→ *Discovery* makes the slow four-mile journey to Pad 39B along the crawlerway. 2005

SECTION 2
ROAD TO THE CAPE

On May 11, 1949, President Truman signed the bill providing for the Joint Long Range Proving Ground, facilities that would have a variety of names over the years, but which would pass into rocket history as simply the Cape.

PROTOTYPE

3

THE GERMAN DEVELOPMENT of rockets during World War II came as a surprise, even to specialists in the field. When space popularizer Willy Ley suggested that the Germans might be ahead of the rest of the world in rocket science in his 1944 book *Rockets*, Robert Goddard, America's reclusive rocket scientist, was incensed. Goddard wrote a letter to a book reviewer on June 10, 1944, dismissing the idea that anyone could be ahead of him and his masterpiece, a 442-pound rocket that he developed through his isolated and secretive experiments in the New Mexico desert at Roswell. Seventeen of his flights had gone higher than 1,000 feet. While Goddard was writing his letter, Wernher von Braun was directing a team of 6,000 people at Peenemünde, a secret island base on the coast of the Baltic Sea. Just three months after Goddard wrote his letter, the first V-2 missile fell on Britain; the supersonic 12-ton monster plunged from a height of 50 miles, carrying a one-ton warhead over four times heavier than Goddard's entire rocket.

← Preparation for the first successful launch of a captured V-2 rocket at White Sands Proving Ground. 1946.

←← A Redstone ballistic missile is hoisted into position for launch at White Sands Missile Range. 1958.

THE CUTTING EDGE of rocketry was within the Third Reich, where it stayed until a large portion of the German rocket science operation showed up near the Austrian border and on May 2, 1945, voluntarily surrendered to the U.S. army then advancing across Germany. Von Braun rather cheerfully offered the Americans not only some 500 of his staff, including many of his best scientists and engineers, but a staggering haul of some 300 boxcar loads of notes, equipment, instruments and V-2 rocket components that the group had smuggled out. The goods had been laboriously hidden in a mine shaft and other secret locations close to the southern front, and a stolen train and forged papers had been used to evade the Nazi SS forces. The Nazis had ordered all the V-2 materials destroyed, along with the engineers themselves if necessary, to prevent them from falling into enemy hands. The U.S. army officers could hardly believe their luck.

Von Braun had recently been arrested by the Gestapo for making no secret of his preference for space rather than war. Accused of using army resources toward his dreams of space exploration instead of concentrating on weapon development, the rocket mastermind had been released from prison only by the intervention of highly placed friends. In those desperate days of 1945, the end of the Third Reich was in sight. Von Braun had put it to his team: to whom shall we surrender? Voluntarily handing themselves over to one faction or another might keep them together and allow them to continue their work. Beside the grim Russians, the likely-to-be-resentful French, and the impoverished British, the Americans were the most attractive choice, and the powerful U.S. economy would offer the best chance of further funding.

The incredible war prize of rockets and rocketeers was shipped back to America, successfully "captured" according to the media accounts and some later histories. Von Braun knew that his team, with its hard-earned experience and expertise, was even more important than the material he had provided at such risk. He and his team would teach the Americans.

WERNHER VON BRAUN AND THE VfR

VON BRAUN, A handsome and charismatic leader of visionary genius, had such a passion for rocketry that he agreed to work for the German army when some of his friends and colleagues had resisted turning their talents toward war. He dreamed so fervently of space rocketry that he had accepted a series of compromises, working first to create weapons for the Nazi forces that funded him, then putting his talents into the hands of Germany's former enemies. It would take time before the world

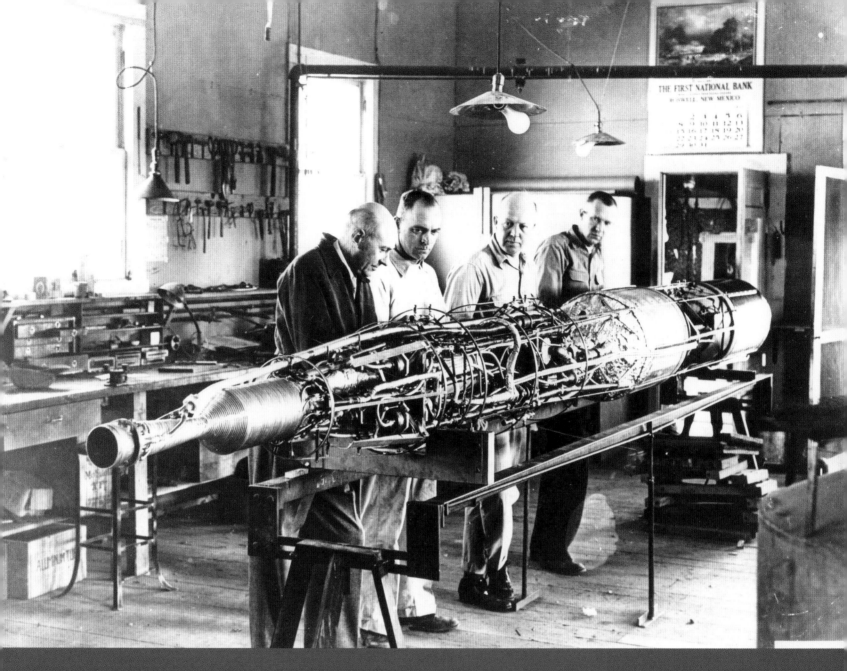

↑ Dr. Robert Goddard, the American rocket pioneer, works on the pumps that inject propellants into the rocket's combustion chamber in his shop in Roswell, New Mexico. 1940.

← Dr. Robert Goddard taught physics at Clark University before leaving Worcester, Massachusetts to pursue his experiments in New Mexico. 1924.

Dr. Wernher von Braun and a group of German rocket pioneers test Herman Oberth's early liquid rocket engine in Germany. 1930.

superpowers were prepared to invest serious money in space exploration, but when that time arrived, von Braun would be ready to play a pivotal role wth his unique abilities and visionary leadership. It had all begun with what seemed like a hobby.

The center of German rocketry in 1931 was a dedicated group of young amateurs who had formed a Berlin club ambitiously called the VfR, or "The Society for Space Travel." The club members built and launched model rockets, proving the principles of rocketry advanced by their president, Romanian theorist and rocket popularizer Hermann Oberth. The club rented the grounds of an abandoned army ammunition depot for its experiments, and with the support of membership dues they were making modest progress.

Yet the VfR soon began to run out of both room and money for their increasingly successful and complex rocket experiments. As the Depression settled in, the military soon became the only viable supporter of continued development, and the following year the leading member of the club signed on with the German army: Wernher von Braun had taken his first step into a larger world.

Von Braun's first rocket demonstration for the army had failed in 1932, but he had nonetheless impressed the military observers. Here was a man who had the ability to inspire others, to lead an effective team, and to balance a thorough knowledge of detailed data with an understanding of the big picture. Pioneering this new field would require a man who could combine solid engineering talent with vision – von Braun was the obvious choice. The commander in charge of the rocket development operation was Artillery Captain, later Major-General, Walter Dornberger, a career military man whose imagination had been fired by the young visionaries. Dornberger selected von Braun to head up the army's rocket unit. The army was interested in rockets because the Treaty of Versailles signed at the end of World War I limited the size of Germany's artillery pieces but the treaty said nothing about rockets.

Von Braun and Dornberger recruited a team and set up shop at an army artillery range at Kummersdorf, in a forest 15 miles south of Berlin. At times their ignition system involved von Braun holding a can of burning gasoline out on a 12-foot wooden pole and tentatively touching it to the nozzle of the latest rocket, which either shot upward or exploded. Yet in a few years the team had progressed so far that in 1937 they launched their third generation rocket, the Assembly No. 3 (or A3) from little Greifswalder Oie island in the Baltic Sea. They had run out of room again.

These technical advancements had won the support of the military, and funding would now be forthcoming for whatever was needed. Dornberger set out to find the ultimate German location for a rocket test range that would provide the ideal setting for maximum development. Von Braun himself suggested the site that was finally selected. In time a massive rocket research and development center was constructed outside a small village called Peenemünde.

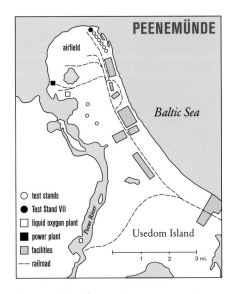

Peenemünde was Germany's rocket development site on the Baltic Sea. The criteria for the selection of Peenemünde were similar to those used for the selection of the Cape in Florida years later.

PEENEMÜNDE

DORNBERGER'S SELECTION CRITERIA that made the Peenemünde location desirable as a rocket proving ground were the identical factors that were later used to select Cape Canaveral as America's prime launch site. In many ways, Peenemünde formed a remarkably complete prototype of the Cape, from its geographic situation to the way launch complexes were designed, equipped and manned, even with some of the same personnel involved. We see the influence of Peenemünde on the launch pads at Cape Canaveral and on America's space facilities.

Peenemünde was sited on the coast, with overwater downrange of about 200 miles, allowing large rockets to be launched without fear of damage to people or property. The rocket center lay on the island of Usedom, its main launch complex located close to the water. Downrange, east of the launch base coastal and island locations lay more or less parallel to the planned course of the rockets, providing a suitable tracking range where a variety of radio, Doppler radar, and camera instruments

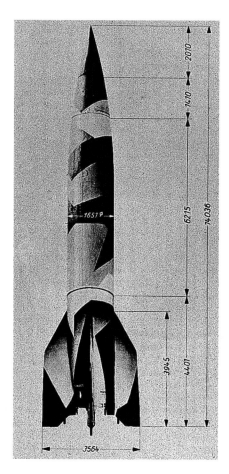

This drawing shows the vital dimensions of the German V2. 1940.

→ **A V-2 missile being readied for launch by the German army at Cuxhaven, Germany. 1944.**

could monitor a rocket's entire flight path. Usedom also offered a fairly flat landscape that could accommodate an airstrip. Finally, Peenemünde was remote. The location was considered a "sportsman's paradise," forested and largely uninhabited. Locating the rocket facility far from population centers provided both safety and secrecy. Every one of Dornberger's criteria for Peenemünde would eventually apply at Cape Canaveral as well.

The rocket facilities turned the village of Peenemünde into a facility the size of a city. Housing, dormitories, workshops, laboratories and a power plant crowded together, providing von Braun with a centralized concentration of the expertise and facilities he needed to defeat the powerful forces of nature that opposed the successful functioning of any rocket. At the north end of the facility lay Test Stand VII, the nest where the powerful new rocket would be fledged.

A factory-like assembly building housed the construction of the missile and its components and the laborious checking and re-checking of the delicate systems of the machine had been integrated successfully. When at last it was ready for test firing, the missile was mounted vertically and secured in a static test tower that enclosed the rocket's nose and left the fins below exposed. The entire tower rolled out of the assembly building into the cold air of the Baltic coast and moved on rails the short distance into the Test Stand, an oval arena ringed by an earthen wall made from debris cleared from the site. The amphitheater-like walls offered protection from the Baltic winds and would help contain catastrophic explosions on the pad. The static test tower moved into place over a flame trench, a large channel built to take the rocket blast safely. This setup was similar to that used at the Cape today.

A4 LAUNCH ATTEMPT: DO OR DIE FOR THE ROCKET TEAM

By October 1942 von Braun's massive organization had produced a flight-ready A4, the rocket that Hitler's propagandists would later name the Vengeance Weapon Two (V-2). On October 3 the rocket was set up on a small launch stand at the heart of Test Stand VII and surrounded by a box-like tower that provided work platforms at several levels. This gantry was enclosed for protection from the Baltic coast cold, described by propulsion engineer Konrad Dannenberg as "nine months of winter and three months of no summer." The rocket was filled with alcohol fuel, as well as hydrogen peroxide to run its pumps. Near the very end of preparations, the cryogenic liquid oxygen was pumped in. Despite the glass and wool insulation around the LOX tank, frost soon coated the hull from

Although it became the foundation of space exploration, the V-2 was developed as a lethal weapon. A V-2 missile landed on Smithfield Market in London, killing 110 people and injuring more than 300 on March 8, 1945.

the ultra-cold liquified gas. Finally, the gantry withdrew on rails to stand clear of the coming blast. Launch personnel were cleared from the area. The operating crew, including pipe-smoking engine development leader Walter Thiel, worked at their instruments in the small control bunker buried inside the earthen berm. Von Braun, Dornberger, and others stood atop nearby buildings for a view of the event.

The rocket was painted in high-contrast black and white to give it maximum visibility for visual tracking and filming. This same paint scheme would be used on rockets built by von Braun's team all the way up through the Apollo Moon missions — but this one would be the first large rocket to fly. The rocket stood on a framework stand almost five feet high and seven feet square. The stand was built around a metal pyramid that would deflect the rocket's flame out to the sides to keep its force from rebounding back into the engine. A mast alongside the rocket carried control hookups and power cables to run the onboard electrical systems until the last second, saving the onboard batteries for flight. The long, methodical countdown had passed smoothly through all the operations and checklists. Three words in sequence would now escalate this situation into the A4's very first launch…if everything worked.

The prototype on the stand was fabricated using Europe's highest technology and its finest precision manufacturing, but the only "instrument" that could analyze the split-second developments during the ignition sequence to determine whether to advance to the next stage was a man named Helmut Zoike. Zoike was the chief of the A4 Propulsion Development Group under Konrad Dannenberg, and he was also the test conductor who supervised static tests and made the final call to launch or not. In the next few moments, relying on his eyes and his experience with static firings at Test Stand VII, Zoike would command this rocket to launch or stall it on the ground as a misfire.

"3…2…1…" and Zoike called for ignition, "*Zuendung*!" A "Christmas tree" burst into sparkling flame inside the rocket engine. The so-called Christmas tree was a spinning pyrotechnic starter, the match that would ignite the propellants now draining from the tanks above into the combustion chamber. If there was a fuel flow problem, the launch team would only get "sparklers" from the igniter, with no flame to follow. But the fuel was flowing, and the Christmas tree had worked: a blowtorch flame was shooting out of the tail of the rocket as it sat on its stand. The rocket wasn't going anywhere yet: this was "pre-stage" ignition, a low-power flame serving only to demonstrate that the engine was working. Zoike, peering through his periscope, would determine whether the combustion was stable within the next ten seconds. The color would tell him whether the propellant mix was right. The strength and consistency of the flame would assure him that the plumbing was working properly, that the fuel flow was strong, and that ignition was uniform and complete. If Zoike didn't like the look of what he saw, he would call the launch off. But this bright, yellow-orange flame looked good; they had a successful pre-stage. "*Vorstufe*," Zoike declared. This was the warning to "stand by for launch command."

Far more than just a suspenseful experiment, this launch represented a potentially life-or-death proposition for those involved. The two previous A4 launch attempts had both failed, with the first rocket cartwheeling into the Baltic less than a mile away, and the second exploding seven miles up. After all the monumental expenditures toward war rocket development—a staggering $3 billion Reichsmarks that could have been spent on valuable practical defenses and vital Panzer divisions—high-level patience with von Braun's rocket game had run dry. If this A4 failed, the group had been told, the project would be shut down. The team would be disbanded and shipped to the Russian front. This was Nazi Germany three years into the war: such threats carried grave conviction.

The test conductor's decision had to be perfect. Zoike had worked

General Walter Dornberger (left) congratulates members of the rocket team at Peenemünde for their successful A4 test flight. Dr. von Braun is second from left in last row. 1942.

with rockets for three years at the Kummersdorf site near Berlin, with four years firing engines here at Test Stand VII. His position reflected the honor of the great trust placed in him. In this critical moment his judgment and perception could be relied upon without question. This was the nature of rocket science: that fortunes and vast efforts of thousands of people could come down to single history-making decisions at moments like this.

Zoike cast the die and gave the command for "main-stage" – "*t!*" His colleague at the firing console threw a switch, dropping the umbilical cable and gunning the rocket's turbopumps. Fuel and liquid oxygen now barreled into the engine as if from firehoses, the mixture becoming a searing fire of thundering, focused force. The blowtorch flame at the rocket's tail burst into an earsplitting dragon's roar. Within a second, the giant pointed tank with fins rose off its pedestal and upward from the amphitheater of Test Stand VII, slowly at first but then with gathering speed. The test conductor felt elation and satisfaction at the sight. It was a true, clean liftoff.

Konrad Dannenberg watched it from the roof of the assembly building just 100 yards away, the rumble of the rocket blast vibrating through

his entire body. The heavy machine was defying gravity straight out, and Dannenberg could feel the raw power resonating in his chest. The A4 rose above the fences and laboratories, above the power plant and the barracks, until it was high above the entire island, heading upward and eastward into the beautiful sunny sky over the Baltic, steering with graphite control vanes buried in the blast flame. It would break the sound barrier without incident and hit the incredible height of sixty miles: the edge of space. The launch was little short of a miracle, and one of the great portents of the 20th century.

The team's wild delight and celebration that day was mixed with awe at what had been accomplished. At a party that evening, Walter Dornberger reminded his team that much hard work remained, and that, for now, they had to focus on the military objective and complete the job of producing a war rocket. He also observed that what they had accomplished was nothing less than mankind's first step into space. No one could foresee exactly where this development would lead. Certainly none of the team imagined that the trajectory of their A4 would lead many of them into the deserts of the American West.

4

LOCATION

THE NEW MEXICO desert stretched out under an early evening sky, the mountains standing high on the horizon like a painted backdrop of raw stone. Horned lizards and rattlesnakes searched for prey among the agaves, while tarantulas and black widow spiders crawled through the waste. Within this setting, a team of engineers, scientists and military men were making preparations. It was early summer in 1947, and the group was about to launch the world's most advanced rocket, a rebuilt German World War II V-2 missile of the same kind that had bludgeoned London just two years earlier. Here in this lonely landscape, under tight security, the U.S. army was now devoting intense study to the weapons of its former enemy, quietly working overtime to learn and absorb the technology of future warfare.

← This view from a satellite in space clearly shows the Cape on Florida's east coast jutting into the Atlantic.

For just over a year now, Wernher von Braun and his team had been voluntarily working for the United States, launching the patched-together leftovers of Germany's final Vengeance Weapons at White Sands, patiently teaching their American colleagues the labyrinthine demands of rocket engineering.

The White Sands team put the V-2s together in a hangar that stood in for their Peenemünde assembly building. Many of the intricate parts had been damaged in shipment, had deteriorated, or were missing entirely from the inventory of hastily stolen V-2 parts. General Electric and other American manufacturers were kept busy manufacturing replacement puzzle pieces, learning rocket science piecemeal as one by one the rockets were completed. The German-American team extended the rocket's capabilities with modifications, using the payload bay not for warheads, but to hold high-altitude scientific research and experimental propulsion system tests. The rocket sitting on the pad this early summer evening was topped with a trial ramjet stage with stubby square wings sticking out on each side like a bowtie on the rocket's nose.

The V-2 had been designed as a field-launched weapon, so it required fairly modest ground support. The rocket was hauled to the launch site on a purpose-built German setup truck called the Meilerwagen. The truck carried the rocket on a cradle that could tilt upward like the back of a dump truck to set the rocket on its launch stand. The stand sat on the concrete launch pad at the Army Launch Area 1, one of four pads laid out in the early days at White Sands, America's first major rocket launch facility. Unseen networks of three-inch conduit ran underground providing command and instrumentation connections to the rocket and the pad systems.

The firing room was a special facility of a new kind necessary for these complicated and dangerous machines. The army had built a heavy-duty reinforced concrete blockhouse near the pad to protect the firing controls, instruments, communications monitors and the launch personnel. The building was built like a bunker with 10-foot thick walls, three blast-proof glass windows, and a concrete roof, 27-feet-thick at its apex, designed to withstand the impact of a V-2 plunging down at 2,000 mph. A washdown system in the roof would decontaminate the building if volatile liquids were expelled during a crash.

A 63-foot-tall service gantry rolled into place on a rail track to surround a V-2 once the rocket was on the pad. The gantry provided a sturdy framework for technician access, with three pairs of work platforms that lowered into place to surround the rocket. Three hoists could lift up to 15 tons of equipment, more weight than an entire loaded V-2. The well-equipped gantry included firefighting equipment, pressurized air hookups and propellant handling gear, as well as electrical and communications connections.

→ German and American scientists working together on the fueling and preparation of a V-2 rocket prior to a test launch at White Sands, New Mexico. 1946.

Filling the big rockets with tons of propellant liquids brought about a weird transformation that still attends such large rockets, a phenomenon attested to by many eyewitnesses from engineers to astronauts who have worked with these monsters at close range. The V-2s were transported dry, for safety and ease of carriage. Unfueled, the rocket was simply a 46-foot tall pair of empty tanks with complicated machinery at the bottom and delicate instruments at the top. It was an inert piece of equipment only slightly more interesting than a big propane tank because of the addition of steel fins. But when they filled this thing with fuel, high-pressure air from the compressors and tons of cryogenic liquid oxygen, the metal began to creak. Hoses hissed. The hull seemed to quiver. The preparations transformed the V-2 into something with a strange presence. It felt alive.

A loaded V-2 contained the equivalent of two swimming pools of alcohol and liquid oxygen, which in combination would burn so fiercely that under control they had the power to send the whole thing 70 miles up into the sky. If that control failed, if there was a leak or a ruptured hose, there was a very large firebomb sitting on the pad. This fact alone commanded respect and fear from the men around the machine—but there was something more to the presence of a "live" rocket. It was like being near a lion or a cobra. The sorcery of science had conjured up some strange and powerful immanence, a ghost in the machine.

RUNAWAY ROCKET

A S THE COUNTDOWN proceeded, electric drive motors slowly withdrew the gantry to a safe distance away from the pad, 500 feet down its rail track, leaving only an electrical power hookup mast standing connected to the rocket. At 7:35 p.m. on May 29, 1947, the reconstituted and modified V-2 recorded as "Missile O" was ready for launch. On command, its engine flared into action. At first it churned out a modest amount of thrust fire, until the control engineers in the blockhouse were certain that the engine was working. Then at their electronic signal the turbopump kicked in, and the blast rose to an amplified blowtorch roar so loud it hurt the ears. The monster lifted off the pad into the air, gaining speed. The White Sands network tracked its rise, and then noticed a guidance system malfunction. A gyroscope—one of the more sensitive and delicate of the V-2's instruments—had failed. The missile heeled over and went wayward, heading not north, but south.

Colonel Harold Turner, the American officer in charge of the operation, expected that the rocket's fuel supply would be electronically cut off to drop the V-2 on the range, but instead it kept going. The missile was tearing southward and a situation was rapidly developing. While the team tracking it in the communications annex behind the blockhouse watched the data in astonishment, the V-2's trajectory crossed the boundary of the test range, heading toward El Paso, Texas. The fully fueled missile could go farther than that during its powered flight time of just over one minute. It screamed across the international border and entered the airspace of Mexico. The monster had escaped.

The V-2 came down a mile and a half south of the city of Juarez. When the team from White Sands arrived, they found a smoking crater 30 feet deep and 50 feet wide, littered with torn metal and surrounded by tombstones. The renegade rocket had impacted in a hillside just outside the Tepeyac Cemetery, and, to general amazement and profound relief, not a single person had been hurt or killed and no buildings had been damaged. It was little short of a miracle. The U.S. army would make good on all the claims for restitution, and the international incident had remarkably little diplomatic fallout. One thing was clear: if America was to keep launching rockets this large, it needed a bigger range.

FROM NEW MEXICO TO FLORIDA

EVEN AS THE chastened group at White Sands was sweeping up seared scrap metal after the Juarez impact, the commanding officials were swearing blind to senior officials right up to the President that a wild rocket escape would never happen again. However, Col. Holger Toftoy, Chief of the Research and Development Division of the U.S. Army Ordnance office, had had an idea for a new and more powerful rocket than the V-2. The army rocket team didn't have the funds at this point to design something from scratch, and in the budget priorities of a postwar economy, rocket science was considered prohibitively expensive. Toftoy thought up a way to make a new rocket out of what they had on hand.

America's rocket development had not won the support of the military as it had in Germany, and outside of Goddard's experiments, American efforts had been modest. The most promising were those of Theodore von Karman and Frank Malina, who had established the Jet Propulsion Laboratories at Cal Tech for rocket research, funded privately, like Goddard's work, by the Guggenheim Foundation. In 1944 von Karman and Malina had designed a little rocket called the "WAC" Corporal, a working scale model of a proposed larger rocket. Compared to the full-sized Corporal the scale model was short and slim, and so it was wryly given its name, WAC being the Women's Army Corps. The first White Sands launch pad had been built for their WAC Corporal, whose Tiny

Tim "starter stage" launch had inaugurated the New Mexico facility two years earlier on September 26, 1945. The liquid-fueled WAC Corporal with its gunpowder Tiny Tim booster stood 25 feet tall. It could achieve altitudes of 42 miles, which wasn't bad, but the rocket was comparatively small and couldn't carry much instrumentation.

Toftoy's idea was that they could stick one of Frank Malina's WAC Corporal rockets on top of a V-2, using the 46-foot V-2 as the WAC's starter stage, instead of the little 5-foot Tiny Tim booster. An entire V-2 ought to give the little thing one hell of a start. JPL would design the WAC-to-V-2 integration system and Douglas Aircraft would build it.

Toftoy's concept worked. By 1949 they were getting the bugs worked out and launching "Bumper" No. 5. The launch was a spectacular success and the little WAC Corporal was hurled upward 248 miles, well into the airless vacuum of space and right off the radar. It would be a year before they could even find the smashed body of the rocket out in the wilderness.

Given that the White Sands Missile Range was only 125 miles long, 248 miles was an especially impressive figure. Looking ahead, Bumpers 7 and 8 were to have especially long, flattened flight profiles. There was no way they were going to fit within that range. Bumper 6 was launched in April, but it was the last one from White Sands. It was time to find a place that could handle the next generation of rocket power.

SITE SELECTION OF THE JOINT LONG RANGE PROVING GROUND

SEVERAL LOCATIONS WERE considered for the "Joint Long Range Proving Ground" envisioned for further rocket development. A leading factor was what lay to the east of the site, since, in aiming for space, you usually want to take advantage of the Earth's spin and shoot eastward in the direction the ground is already moving. Going with the spin gives you a strong head start on orbital speed. As soon as members of the selection committee understood that the rocket path would be toward the east of the launch site, locations with unsuitable downranges started getting crossed off the short list. Texas was very keen to win the selection, but the downrange was a state called Mississippi. Southern California's downrange was Mexico...again.

The most promising site was an abandoned naval station in Florida, sited on a landmark called Cape Canaveral. The Cape was a sleepy shrimping and fishing spot dotted with a few farms. The area was a hazard to navigation with its treacherous shoals, and its point was marked by an 1847 lighthouse. The palmetto wilderness was tangled and primitive. The

Spanish explorers and conquistadors named it Cabo de Canaveral, "the Cape of the Canes," after its marshes filled with thick reeds.

The military analysts examined the location. First and foremost, no rocket was going to run out of room here. The Cape covered about 15,000 acres, and you could stand on the Cape shore and look out over a 5,000-mile shooting range stretching across the Atlantic. The location also offered the very same situation that had attracted Dornberger to Peenemünde: the flat landscape was suitable for airstrips; the large overwater downrange was dotted with islands on which to base tracking stations; the Cape was a thinly populated region, yet within reasonably easy reach of modern transportation.

The selection group settled on the Cape in 1947. On May 11, 1949, President Truman signed the bill providing for the facility, and in 1950 the Cape formally became the Joint Long Range Proving Ground, the genesis of the facilities that would have a variety of military names over the years, but which would pass into rocket history as simply the Cape.

FIRST CAPE LAUNCH

"**D**ON'T STOP OR you'll bog down," they would tell people trying to drive out to the rocket site. The three roads leading through the palmetto scrub to the first Cape launch pad had been made passable—more or less. Travelers could still sink into the muck if they weren't careful. All the launch equipment and the rocket itself had to be brought in by these dubious routes.

On July 24, 1950, a black and white needle-tipped Bumper V-2 stood on a launch pedestal in the middle of a 100-foot-wide concrete apron called Pad 3. Of the four launch sites initially surveyed, Pad 3 was the highest and driest in the wet sandy ground, so, in the rush to activate the Cape, it was the first one built. The site wasn't yet approved for "beneficial occupancy," but it would do for the time. The entire U.S. Army Corps of Engineers team dispatched to build the first Cape launch complex had been involved personally in the pad construction: that is, all three of the Corps men had helped their contractor mix and pour the concrete. The ground had been cleared and leveled for the pad just 45 days earlier.

The sun-drenched morning was far from the climate of Peenemünde, and in fact the sweltering humid atmosphere here had unusual properties. It was composed largely of mosquitoes. Between the broiling heat, the lack of air conditioning, and a plague of flying blood-sucking pests, it took real dedication for the launch crew to maintain the necessary focus and energy to carry out their work.

As the frying climate took its toll on the men, it had already taken out one of the rockets. Bumper 7 had basically fainted a few days previous, a fizzling misfire on the pad due to salt air corrosion in a fuel valve. The engineers would have to scrub at it for a few days before they could try it again. The press was dubiously gathered to see whether the next rocket, Bumper 8, would work any better.

The setup was not especially convincing. The standard V-2 ground equipment was on hand—the German rocket setup truck, a tanker carrying liquid oxygen, and drums of chemicals to fuel the WAC, but instead of White Sands' stout and elaborate service structure, the gantry here was a 55-foot tower of painter's scaffolding with plywood work platforms from Orlando. They had it on casters so they could push it around as necessary, and, if more than ten engineers climbed aboard, it swayed.

The control center looked like a miniature version of the blockhouse at White Sands, but instead of a 27-foot-thick concrete roof, this shack had three-quarter-inch plywood. It was hunkered down behind a sandbagged berm, and, for added protection from shards of searing hot metal expelled at ballistic speeds, the plywood was covered with one layer of tarpaper. The control shack stood just 300 feet from the pad. Civilian emergency officer Norris C. Gray believed that the best protection this lemonade stand offered was "shade from the sun." Gray had seen the V-2 craters out at White Sands, and, in his estimation, if the rocket exploded a second or two after liftoff, when it was above the berm, they'd have a clean slate to work with at Launch Pad 3. To see the rocket from inside the control house, one looked through the window into two angled bathroom mirrors set one above the other like a kid's periscope.

The concrete pad looked plain but hid more sophistication than met the eye. Underground walk-through passages carried conduits and high-pressure air under the pad to the launch point at the center. A built-in system at the surface wept water to help keep the concrete from spalling at the launch blast. The pad itself was carefully sloped to control the runoff of water and propellant liquids. Gray had introduced the idea of directing the different kinds of runoff into stainless steel troughs leading to specific collection pans to keep the ground uncontaminated by the various caustic liquids in use. Hidden conduits carried cables below the surface. Good thought had gone into making the best use of available resources at the time. Nonetheless, the Cape was not yet up to the standard of White Sands, and it was puny compared to Peenemünde's mighty Test Stand VII. However, the Americans couldn't have wayward rockets landing on any more graveyards, let alone cities, so they were going to make this place work.

Bumper 8 roared to life, shooting Cape sand in all directions. It soared into the sky over the palmettos before leaning over to head out

Cameramen film the Bumper V-2 liftoff, the first missile launched at Cape Canaveral. July 24, 1950.

to sea. Its mission was to successfully launch its second stage while flying almost horizontally. It didn't work. The second stage failed. The falling V-2 was deliberately blown apart 48 miles out to sea to make sure it didn't do any damage. Debris rained down on the ocean harmlessly and that was the end of Bumper 8.

The launch crew held a party of sorts that night, but everyone was physically, mentally and emotionally drained by the experience and many of them simply hit their bunks exhausted. Incredibly hard work: that's what it took to launch rockets. It took bold attempts, and carefully studied failures to make progress.

In the years ahead the Cape would see triumph and tragedy, but it would never see surrender. The first rocket had taken flight, and more would follow. The cornerstone was laid, upon it would rise America's Gateway to Space.

SECTION 3

RACE FOR THE HEAVENS

The rocket will free man from his remaining chains, the chains of gravity which still tie him to this planet. It will open to him the gates of heaven.

– Wernher von Braun

THE CHALLENGE

5

THREE YEARS AFTER the makeshift Bumper launch pad, the Cape was beginning to look like a proper rocket launch facility. The first of the great launch complexes were a pair of pads numbered 5 and 6, built under the supervision of Wernher von Braun's trusted launch operations supervisor, Dr. Kurt Debus.

Von Braun's charisma and energy tended to overshadow those around him, but Debus was another of the team's powerful assets; behind his mild-mannered demeanor the man possessed hidden talent...and hidden drama. The gouges decorating his chin were fashionable fencing scars that he had proudly won in his youth in elite combat with other swordsmen back in Germany. Debus handled rocket engineering as skillfully and precisely as he had once handled a sword, and his expert judgment was completely trusted by von Braun. On the job, Debus was a supportive and considerate leader, but he demanded the highest standards of dedication and expected a serious demeanor from his men. The typical casual attitude of the Americans did not, in his eyes, mix well with the demands of rocketry, and he expected his team to take their work very seriously—doing the job well was not enough.

← The third Mercury rocket, carrying pilot Alan Shepard, lifts off from Cape Canaveral. 1961.

←← The Original Seven Mercury astronauts. From left to right: Scott Carpenter, Gordon Cooper, John Glenn, Virgil "Gus" Grissom, Walter Schirra, Allan Shepard and Donald Slayton. 1963.

↑ Mercury Redstone prelaunch activity. 1961.

→ Dr. Kurt Debus and Dr. von Braun relax at Launch Pad 12. 1962.

WITH THE OUTBREAK of the Korean War in 1950, the U.S. Army asked von Braun to build a new, American rocket. It was about time, since they had pretty much run through the leftover Nazi V-2s. The Germans, who felt they had been doing nothing for almost ten years, were gratified to finally get to build something innovative again. Stationed now at the Redstone Arsenal in Huntsville, Alabama, the team devised an American version of their V-2, named, after its place of origin, the Redstone.

Bernard Tessman had engineered the great engine test stands at Peenemünde. The rocket team was now static-firing the Redstone on Tessman's new stands in Huntsville. The arsenal offered enough room for artillery testing, but to actually launch the 200-mile-range missile von Braun would need the Cape and proper facilities there. Now that the Army was on board because the Redstone was a field weapon, there would be no shortage of money, and the facilities would be first-class. Such is the budget distinction between science and war in a dangerous world.

Debus was dispatched to the Cape to supervise the design of the ideal rocket launch complex. Sites 5 and 6 would be the prototype of all future space complexes at the Cape. There would also be twin pads, as the Germans liked to have backups. Malfunctioning rockets tended to explode and explosions damaged pads: backups kept the schedule on track in spite of setbacks. So there would be two pads, and they would share the expensive control facilities that were less likely to be damaged in a pad accident. Debus and his team worked up a bunker just as impervious to explosions as the old blockhouse at White Sands and the small control room buried under Peenemünde had been, although it was much better equipped than either of its ancestors. The Redstone blockhouse would be built like a fortified pillbox, with heavy, square lines and slot windows that gave direct views to each pad through thick armored glass. Crowded inside were controls and instruments: voltage meters, circuit selectors, temperature warning lights for the computer and terminal equipment, bulky switchboard-type guidance computers and reel-to-reel tape machines.

A big scale told the launch crew how much the Redstone and its fuel weighed. If a glitch occurred during countdown, the crew would have to work the problem until they fixed it…meanwhile, to prevent its tanks from bursting, the Redstone would be venting boiled off liquid oxygen from a pressure relief valve. A long enough delay could deplete the oxygen supply and leave too little aboard for the flight, so a tanker truck would have to go out to top off the tank. When the scale in the blockhouse showed the correct weight, the crew, dressed in their bulky safety gear, were given the high sign and left the pad in a bit of a hurry.

Within the futuristic, purpose-built consoles, which foreshadowed

A U.S. Army Redstone rocket, developed as a ballistic missile under the direction of Dr. von Braun, is being lifted into position for launch at Cape Canaveral. 1953.

those of *Star Trek's U.S.S. Enterprise* bridge, there were pullout miniature cylindrical drawers. Today, one of them still holds its original contents: cigarette butts. In the culture of the 1950s, any activity as stressful as rocketry was accompanied by a great deal of smoking. Ashtrays were considered such a vital component of the control consoles that they were built in and given a clean futuristic look like everything else.

The concrete pad and the small launch pedestal with flame deflector closely echoed the simple arrangement of the Bumper setup at Pad 3. The new launch gantry tower was a state-of-the-art facility that gave technicians easy access to every part of the rocket. It was patterned after an open-faced oil derrick, and had been built across the country in Oakland, California, then shipped to the Cape by rail. The Redstone gantry became the prototype for the later Cape gantries that sprouted all along the coast throughout the 1960s.

Complex 5/6 racked up almost thirty launches between 1956 and 1961, winning von Braun's rocket the nickname "Reliable Redstone" for its solid performance. The short-range Redstone gave the U.S. Army its first "battlefield" ballistic missile. However, strategic needs were changing,

At a special ceremony in Huntsville, Alabama, Dr. von Braun and 102 other German scientists and engineers, along with their families, take the oath of citizenship to become United States citizens. 1955.

and, behind the Iron Curtain, the Soviet Union was developing rockets that would reach far beyond the battlefield, all the way into space.

THOR

By 1955 intelligence regarding Soviet advances in missile development had U.S. leaders worried. The United States had no medium-range or long-range missile capability and, after a decade of letting von Braun's team languish in idleness, it was finally decided that it was necessary to develop a strategic nuclear missile as quickly as possible to ensure that the Russians would not have one first. The air force would be in charge of building a medium-range rocket with a 1,500-mile reach, so limited that it would have to be based in Britain to threaten the USSR. It wouldn't be very powerful, but at least it would be something and, as a simpler system, it could be available before any of the long-range alternatives then under consideration. Limited range notwithstanding, the missile would carry an atomic warhead, and it took the name "Thor" after the Norse god of thunder. On December 1st, 1955, President Eisenhower gave the Thor missile "highest national priority," and initiated one of the most extreme crash development programs in the annals of aerospace.

Thor led to the construction of what was becoming the standard suite of facilities for new vehicles at Cape Canaveral: alphabet hangar M was used for assembly of the missile, and a launch complex, LC-17, was prepared to send it off. Complex 17 would have a blockhouse, a service tower, and a pair of pads, A and B. As at the Redstone complex, the twin pads were a nod to the likelihood of trouble ahead.

The Douglas Aircraft engineers on the Thor project had a critical head start: a discarded rocket engine from the Atlas missile project. Douglas built its Thor rocket around this cast-off engine in just thirteen months, accomplishing one of the fastest rocket developments in history. Douglas' team had their first launch-ready Thor sitting on Pad 17B at the Cape on January 25, 1957. Their amazing production speed almost made it look easy.

Rocketry is never easy. The prototype Thor 101 ignited for launch and the engine roared to life, lifting the new bird into the air for just moments before a liquid oxygen valve failed. The Thor lost thrust and dropped back to the pad, falling right through its launch ring and exploding. The brand-new pad lay smoking, blackened, and badly damaged.

Backup Pad A was six months away from completion. The Cape crew had no choice but to wade into the grime and repair Pad B for another launch attempt. The second Thor did better, but the third launch attempt in May didn't even make it off the pad. Five minutes before ignition, the rocket's fuel tank ruptured and another horrible fireball vomited wreckage, kerosene and burning mess all over the complex. By that time Pad A was ready, and the fourth Thor could be sent up while grim Pad B was refurbished again. Pad A didn't stay clean for long. On October 3, the sixth Thor failed after liftoff and dropped through the launch ring for another pad-wrecking detonation.

Cutting-edge rocket engineering is always a battle against the near-impossible. Thor's carefully studied failures provided valuable information each time, and the system was improved until it was beginning to work through its proving-launch schedule by the end of 1957.

In 1958 the air force would set up a Thor launch crew training site on the California coast north of Los Angeles, at a location that would become Vandenberg Air Force Base. In 1959 the air force would deploy Thor missiles in Great Britain, but the Thor stopgap came much too late to match the stunning debut of the Soviet R-7 ballistic missile. In October 1957, while the 1,500-mile stopgap was still exploding on the pad, it was an R-7 that orbited a little satellite called *Sputnik* and changed the world.

→ **A Vanguard missile carrying a test satellite explodes in a ball of flame after rising only a few feet off the pad.**

VANGUARD

The U.S. counterpart to *Sputnik* sat at the Cape on Pad 18A, on December 6, 1957. It hadn't exactly been made welcome by Cape military authorities, who considered scientific programs a critical waste of time and resources in the face of urgent missile development needs. The scientists had to elbow in at a spare pad site. The rocket was a delicate-looking, three-stage experiment called Vanguard. The entire load this slender vehicle had a chance of putting into orbit sat in its nose looking like a chrome grapefruit with antennae.

Wernher von Braun had been champing at the bit for several years to loft a satellite with the rugged Redstone, but the Department of Defense had rejected von Braun's proposed Project Orbiter in 1955, and instead awarded the honor of launching the first American satellite to the Naval Research Laboratory. President Eisenhower was dubious of space exploration and didn't want pointless scientific adventures interfering with military missile development. The Vanguard rocket was so weak that it had no value for military applications. The project was intended to be the highlight of the International Geophysical Year of 1957–1958, a worldwide coordination of scientific endeavors. With the surprise debut of *Sputnik*, public imagination suddenly required Vanguard to serve as a riposte to the Soviets, a spotlight role the experimental scientific rocket was never intended to play.

On the momentous December day, Vanguard rose for two seconds off Pad 18A before its unstable thrust dropped it back onto the pad. American pride went up in flames as it exploded into a rich, billowing red-orange fireball live on national television. Kerosene, nitric acid and hydrazine from the rocket's multiple stages erupted in every direction. You could practically hear the Russians laughing in Moscow as across the country hearts sank, stomachs churned and faces burned with shame. Vanguard's nose cone and satellite toppled off the sinking fuselage as the rocket collapsed into its pyre of disintegration. The American press dubbed it "Kaputnik."

Von Braun would get his chance after all.

EXPLORER

Von Braun's team had added complex solid rocket upper stages to their reliable Redstone to create a "Redstone plus" called the Jupiter C. More powerful than Thor, the Jupiter C was intended to send a one-ton nuclear warhead 1,850 miles. Kurt Debus' team at the Cape had prepared a pad for the Jupiter C next door to the Redstone pads. So many rocket and missile systems had already been developed that the Jupiter pad got registry number 26.

Von Braun had long known that his Jupiter C had the power to put a small satellite into orbit, but his team had been ordered to ballast the Jupiter C upper stages with sand during tests to make sure that they didn't "accidentally" get into orbit before the official Vanguard program gained that honor. The national embarrassment of *Sputnik* compounded by the humiliation of the Vanguard failure had finally given von Braun's group freedom to do what they had wanted to do since the days of World War II. The team worked hard to ready a Jupiter C for its great mission. For this they gave it a new name: Juno.

As soon as *Sputnik* had gone up, the Jet Propulsion Laboratories at Cal Tech had been given orders to prepare a satellite for von Braun's rocket—as a backup in case Vanguard failed. JPL designed and built a 31-pound satellite in 84 days. The pencil-shaped Explorer I was stuck right on top of the rocket like a needle, unprotected by any shield or enclosing nose cone.

Explorer I blasted into the heavens on January 31, 1958 and gave America its first satellite in space. The Geiger counter carried by the probe allowed scientist James Van Allen to discover the great bands of radiation surrounding the Earth, now known as the Van Allen belts. Explorer I's launch was another milestone at the Cape, the herald of many orbital triumphs to come. The Jupiter/Juno would go on to launch more satellites, but everyone knew the Space Race was on, and it was only a matter of time before a man rode a rocket into space.

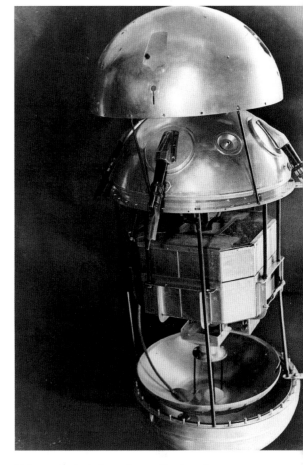

This exploded view shows the interior of the Soviet *Sputnik 1*, the first object successfully launched into earth's orbit. 1957.

6

MERCURY

S O MANY ROCKET explosions had scarred the Cape and its skies. How could we possibly put a man aboard one of these flaming instruments of destruction? NASA selected seven men who knew the risks and wanted the job anyway: the Original Seven, America's first astronauts. Their first ride would be on the sturdiest rocket available: von Braun's reliable Redstone. Its power fell far short of what it would take to send a man and his spacecraft all the way around the world, but it could hurl him up to the edge of space on an arc that would end in a splashdown a couple of hundred miles out in the Atlantic—if everything worked.

← The monument at KSC honoring Project Mercury features the number 7, representing the seven original astronauts, inside the astronomical symbol for the planet Mercury.

→ Cuban Premier Fidel Castro walks with Russian cosmonaut Major Yuri Gagarin, to a reception in Gagarin's honor as the first man in space. 1961.

↓ Russian Premier Nikita Khrushchev (center, first row) hosts a celebration in Moscow to celebrate the first anniversary of Yuri Gagarin's pioneer space flight. Cosmonaut Gherman Titov is far left, and Gagarin is to Khrushchev's left. 1962.

To survive a trip on a Redstone, a man would need a protective space capsule. McDonnell Aircraft in St. Louis would build these spacecraft, which looked like missile nose cones but packed an unprecedented amount of equipment. The capsule had to hold an astronaut seated on his back so he would not be crushed by the acceleration forces of launch. The spacecraft would have to contain life-support systems and extensive instrumentation. The capsule's nose would carry a parachute to drop it into the ocean in one piece. All its equipment and protective armor had to be wedged into a tight package, and it couldn't weigh more than the Redstone could carry.

The capsule would be only a few feet wide, and outside the hatch, death would be inches from the astronaut's face. A hypersonic slipstream would be right outside during launch. A lethal vacuum would surround him while he floated weightless through the top of his arc in space. Searing heat would envelop the craft during its high-speed plunge back into the atmosphere during re-entry. All this would pass inches away from the astronaut within the space of 15 minutes. These men were going to earn their pay and then some.

The prospect of placing an astronaut on top of such a dangerous machine, closing the door and lighting the fuse sobered the pad community. The astronauts were not just human beings: to many of the crew at the pad they were co-workers and friends. For all the precision, confidence, engineering brilliance and hard work, rocket engineering was still desperately risky when a human life was on the line. At the center of all this complexity and machinery would be a beating heart and warm blood; all of that man's bravery, skill and spirit would not keep life in his body if any of a thousand and one things went wrong. The pad crews knew this; they could feel it. They knew that they would bear direct responsibility for many of those possibilities.

Safety and quality assurance inspectors were added to the group to look over the shoulders of the pad crew. More than ever, the workers at the Cape realized that they formed the end of the line. There was no one else to catch anything after them. The only one left would be the astronaut, and he would not be carefully recording what went wrong, he would be dying in a fireball.

The spacecraft kept getting updates and modifications. Hangar S was busy day and night with astronauts running simulations of the first Mercury mission. A chosen three—Gus Grissom, Alan Shepard and John Glenn—all trained for the flight relentlessly. Shepard alone made some 120 simulated flights during this time.

On April 12, 1961, the first human being made the journey into space, but he was not an American. Soviet cosmonaut Yuri Gagarin soared

The first and second groups of American astronauts pose together for an official photograph. Seated are the Original Seven, selected in 1959 (left to right): Gordon Cooper, Virgil Grissom, Scott Carpenter, Walter Schirra, John Glenn, Alan Shepard and Donald Slayton. Standing are the second group of astronauts, selected in 1962 (left to right): Edward White, James McDivitt, John Young, Elliot See, Charles Conrad, Frank Borman, Neil Armstrong, Thomas Stafford and James Lovell Jr. 1963.

not just into space, but entirely around the Earth, a doubly spectacular leap that ranked with the first flight of the Wright Brothers in its profound historical significance. Gagarin's flight aboard *Vostok I* lasted almost two hours before he returned to land in the Soviet Union.

The bar had been set, and it was much higher than the U.S. could reach. The defeat was crushing to those working on the space program, but there was nothing for it but to focus and keep working as hard as they could.

It was not until May 2 that Shepard was announced to the public as the pilot chosen to fly the first American astronaut launch attempt. Since the Vanguard explosion, the public had known what danger lay in a rocket's power. The community at the Cape knew it all the more. Some of the pad crews had seen over a hundred disastrous failures by the time Alan Shepard's rocket sat on Pad 5. The astronaut's daring in the face of this danger electrified everyone. From the Cape epicenter all across America, people felt excitement and awe at what the astronauts had volunteered to do.

The launch had already been scrubbed for bad weather twice. It was early on May 5, 1961, that preparations began again. By 1 a.m. they had begun loading the tanks with propellants. Gallon after gallon of kerosene poured into the lower tank, and liquid oxygen (LOX) so cold it could burn your skin on contact poured into the upper. Vented oxygen made white clouds that drifted around the launch site like vapors around a witch's cauldron. The spacecraft's thrusters would run on hydrogen peroxide, a fuel von Braun had described for such use to the readers of *Collier's* ten years earlier. The pad crew filled the capsule's peroxide thruster tanks in the predawn.

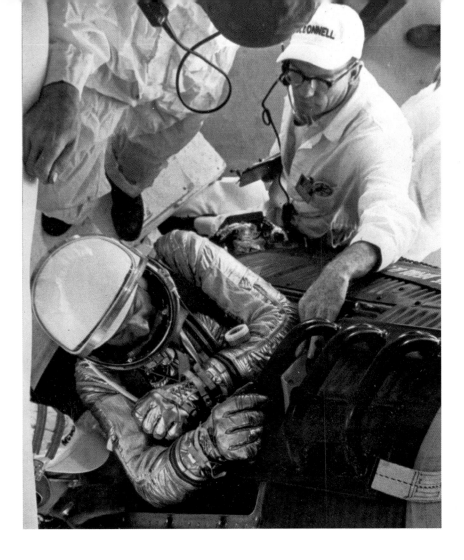

Mercury astronaut Alan Shepard is helped into a Mercury capsule for a flight simulation test. 1961.

A van brought Shepard to Pad 5. It was still dark. He stepped out in his silver suit, carrying his personal air conditioner like a space-age brief-case. Until the capsule's system was hooked in to replace it, the briefcase would keep the tightly-wrapped astronaut breathing and at a comfortable temperature. Shepard waved, ready to keep his date with history.

The Mercury astronauts said that you didn't board a Mercury capsule, you strapped it on. Once upstairs in the clean room, Shepard had to be eased into the tight space available for him. It was like loading someone into a car trunk, except that the galaxy of switches and controls filling the cockpit had to be carefully avoided during boarding. The technicians strapped Shepard in using tight restraints to hold him against the forces he would experience. Oxygen hoses. Communication lines. Medical instrumentation connections. They integrated him into the machine until his pulse showed up on the control boards, just like the rocket ship's fuel pressure and electrical charge levels. This strapping-in was the routine they had practiced until it was second nature — except now it was about to give way to something they had never practiced … Shepard was going to be on board when they launched this rocket.

All seven Project Mercury astronauts at breakfast just before Alan Shepard (center, back row) takes off on Freedom 7, America's first manned sub-orbital flight.

Pad crew leader Guenter Wendt, an ex-Luftwaffe mechanic and friend of many of the astronauts, gave Shepard a final thumbs-up. The realization of what was about to happen to his friend was finally sinking in. Launch control gave the order for closeout, and the technicians fastened the hatch shut with its 70 bolts. Soon the room was cleared, and Shepard was alone. He could only see the outside world through two portholes and a small periscope. He would be nearly blind to the whole experience. The translucent green chrysalis atop the service tower parted. The clean room opened away from the capsule, and then the gantry tower rolled back to give the rocket room for the coming blast.

Was there anything I forgot? The thought gripped many of those who now gathered away from the pad at the fallback area. The weight of responsibility hung palpably in the air.

Above Pad 5 Shepard awaited the final countdown. The cocksure flyboy had jockeyed and toiled in order to get this supreme isolation at the tip of the first American spaceship, and now he had it. And yet, just like the pad techs, he felt the responsibility that focused so squarely on him above all others: this perfectly human superman was hoping intensely that he would not be the one to ruin the mission.

What about things beyond his control? What if a fire broke out at the pad while Shepard was strapped into his cramped capsule? The gantry tower by which he had boarded his rocket had rolled several hundred feet away. The "cherry picker," a yellow crane with a basket at its end, was kept close to the capsule until the very last moments before the launch. If

An early Mercury Redstone being
prepared for launch. 1961.

a pad fire broke out, the basket could be guided by remote-control back up to the capsule side. Shepard would open his hatch, struggle to unstrap himself, scramble out of his capsule and step into the basket without plunging to the concrete below, and then the crane would pull him back out of the way… as long as the fire burned patiently enough for all this to happen in time to save him. This escape route was far from ideal, and as a primary safety system, it would not last more than two Mercury missions. Right now it was all there was.

And if there were problems once he was in the air? There was no ejection seat, although he could bail out of the rocket by taking his whole capsule with him. One of the most important systems on the whole rocket was the bright red escape tower that formed the capsule's needle tip. During the boost phase, while the Redstone rocket was blazing behind him, Shepard, his autopilot, or ground control could all fire the escape system if something went wrong behind him. The escape tower nose rockets would pull Shepard's capsule away from the Redstone (at a potentially back-breaking acceleration of 20 Gs) and then the capsule would fall back to Earth to float down on its parachute to be recovered, just as planned for a normal flight.

Such was the theory, but would this system actually work in an emergency? In May 1961, an unmanned Mercury capsule went up on an Atlas rocket test launch. A critical failure occurred and, hidden inside a cloud, the Atlas exploded. The capsule soon appeared, drifting back safely, thanks to the launch escape tower which pulled it free, just as promised. NASA had planned well for the catastrophes it could imagine, but no astronaut had yet put all the planning to the acid test, to learn whether there were catastrophes that had not been imagined.

The nation's eyes were on the launch. This was a test: who were we? Alan Shepard would tell us about our bravery. NASA would tell us about our capability. Our President would tell us about our potential. The identity of the United States of America was at stake.

Some 350 media correspondents busily streamed out their reports from the Cape. Some leaders in NASA and Congress worried that too much attention to another potential Vanguard debacle would do good for no one but the Russians. Four days before Shepard's launch date, NASA administrator James Webb had issued a caution: "We must keep the perspective that each flight is but one of the many milestones we must pass. Some will completely succeed in every respect, some partially, and some will fail." But the American way was to show the taxpayers what they had paid for; NASA was a civilian agency, and news coverage would remain open, no matter what the consequences.

Shepard's astronaut comrades were all around him, involved in every part of the operation. John Glenn had breakfasted with him and stood in for him during the pre-launch preparations. Gordon Cooper was in the blockhouse with the launch crew. Deke Slayton was on the radio link in Mercury Control nearby. Wally Schirra was circling in his F-106 chase plane, scouting the winds and ready to follow Shepard as high as he could.

The launch team delayed again and again, checking and rechecking everything, waiting for completely perfect readings, everyone fearing to initiate the blast that could kill the man aboard their rocket. Alan Shepard knew that perfect assurance would never arrive. Over the communications loop, he told the launch team in frustration, "Let's light this candle." His courage gave the team the push they needed.

One of the men who was in the blockhouse recalls: "Once the fire button was pushed, it took about five seconds for the fuel tanks to pressurize. Then, when the rocket lit up, we had to wait and watch another 20 seconds while this thing built up enough thrust to lift us off." The Redstone missile carrying Alan Shepard rose from Pad 5 and shrank into the sky.

Shepard streaked through the sky at 5,100 miles per hour, shooting up to a height of 115 miles, firing communication back and forth with Mission Control and demonstrating that he could operate his ship perfectly well despite the initial stress of high acceleration forces and the later weightlessness. On the other side of the peak, he splashed down 15 minutes after launch, 302 miles offshore, his capsule and his body in excellent condition.

His spirit, like that of his country, was more than excellent. As an individual, Al Shepard was an ornery cuss and one tough customer, but his voice sometimes trembled during the next few days when he told the story of those 15 minutes. He was touched by the magnitude of the event. He was human—a man with flaws who could worry about messing up like anyone else—but he had faced the ultimate test with courage and had delivered the goods under pressure, doing his job perfectly when it really counted. He had brushed wings with the angels.

NASA's original plan was to send each of the Original Seven up for a flight on the Redstone. It didn't work out that way because the Russian space program was leaping ahead. While America celebrated its first man in space, Soviet premier Nikita Khrushchev scoffed at Shepard's suborbital hop, calling it "a flea jump" in comparison to Gagarin's orbit. The wisecracks stung because the Russians were in an excellent position to make them. Gus Grissom went up from Pad 5 for his Mercury-Redstone flight in July for a nearly 16-minute mission, but in August the Soviets launched cosmonaut Gherman Titov on an orbital mission lasting an

Alan Shepard dashes across the deck of the aircraft carrier USS. Lake Champlain after his successful suborbital flight in the Freedom 7 Mercury capsule. 1961.

entire day. The Soviets' sneering superiority made American best efforts look pathetic.

In a bid to catch up and close the gap, NASA made the decision to scrap all five of the remaining sub-orbital Mercury-Redstone flights. The program had already told us what we needed to know about basic space-flight operations.

Gus Grissom's Mercury mission was the Redstone pad's 31st and final launch. Complex 5/6 had done its job and put the first Americans into space. The cutting edge of manned U.S. spaceflight would now move a few pads up the coast. We would skip ahead to try to get a man in orbit immediately.

AFTER EXPLORER, VON Braun's rocket team in Huntsville began developing their Redstone and Jupiter systems into a dramatically new vehicle. The Saturn, as it would be called, would take years to design and build. Fortunately, von Braun's was not the only rocket team in America anymore. American companies were now well into fielding bold new rocket ideas of their own. Powerful new rockets stepped into the spaceflight spotlight at the Cape. The first of them was the Atlas.

MERCURY-ATLAS

A DOMED FORTRESS PROTECTED the launch crew from the potential wrath of the Atlas. The igloo at Pad 14 would command the momentous launch of America's first astronaut into orbit, and all those who would follow him riding in Mercury space capsules atop Atlas missiles. The main floor inside the blockhouse was 60 feet in diameter, and there was a second story above it. The protective walls had 10½-foot-thick cores at their base, with a further 40 feet of sand around them to dissipate the force of a catastrophe. A gunite concrete shell held it all in place. The astronaut, of course, would be sitting up there in the little capsule right on top of the rocket. For protection he had a helmet and testosterone.

The new rocket was the Atlas, America's first intercontinental ballistic missile built by Convair for the U.S. air force. This was a long-range rocket with the power to carry a lightweight atomic bomb all the way to the Soviet Union. It had been brought to life by two key engineering elements: a Rocketdyne engine derived from the Navaho winged missile booster engine, and remarkably lightweight tanks.

The Convair engineers had discovered they could make the Atlas tanks a lot lighter than other rocket tanks by stripping out nearly all their structural support and shaving the hull down to the thickness of a dime. It naturally followed that a lighter weight rocket could carry a heavier payload farther. The removal of the structural skeleton, however, made the tanks so feeble that the rocket couldn't even support itself. It could not stand up straight without collapsing, even on the pad. The only way to keep an Atlas standing was to pressurize the tanks: the Atlas was like a big stainless steel balloon that only got its structural integrity from pressure in the tanks. It was a wild idea. This was American ingenuity and daring for sure. The Atlas had leapfrogged the German rocketeers' sturdy "Reliable Redstone," but at the price of becoming a very delicate and unreliable system. The intensity of the Space Race demanded such desperation. In 1962 the Atlas was the only U.S. rocket available that had a chance of putting a man into orbit. Accordingly, the astronauts paid close attention to the Atlas development program. In the middle of their training, the flyboys came to the Cape to watch a test launch of their new ride. The Atlas flared to life and rose into the blue, exhibiting its great power all the way up to and including the point where it folded in on itself and burst into a spectacular fireball. As wreckage rained down over the ocean, the astronauts looked at one another. Four out of ten Atlases were exploding after launch. There was a reason that we sent up monkeys first.

NASA selected John Glenn to pilot the first manned Atlas rocket. The

Chimpanzee Ham and technician go over equipment before Ham's Mercury Redstone test flight. 1961.

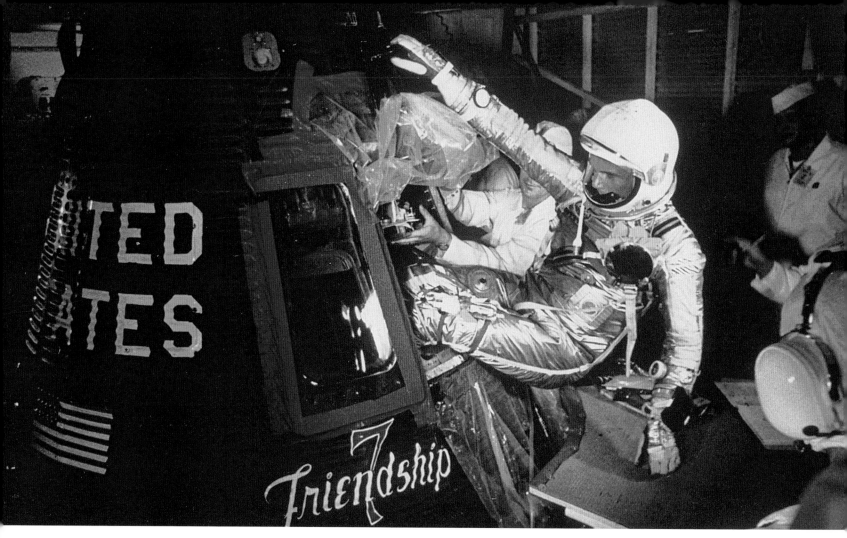

Astronaut John Glenn is helped into the Mercury Freedom 7 capsule to begin his manned earth orbital mission. 1962

pressure on him was even greater than Alan Shepard had faced. The reach was much higher—not just a 15 minute jump for a brush against space, but a journey circling the entire Earth. With this feat America would catch up with the Soviets, at least for the moment. To do this, John Glenn would be riding a volatile missile that represented the leading edge of rocket engineering. This launch would be perhaps the most intense moment in the Space Race.

Col. John Glenn was loaded in and ready on February 20, 1962. Pre-launch procedures included clearing all the crew personnel away from the launch tower before the controller threw the switch and turned on the capsule power system. Why clear everyone away? The pyrotechnics could go off accidentally when the power activated and fired the escape rocket by mistake—or worse. This was the kind of hair-trigger bomb that Glenn was aboard. At T-90 minutes, they bolted the hatch down over Glenn in his seat, but found one of the bolts broken. The count held for 40 minutes while the crew replaced it.

With 35 seconds to go, the tower umbilical connection dropped away from *Friendship 7*, as Glenn had named his ship. The countdown hit zero and the engines ignited. Flame blossomed at the base of the silver Atlas, and thunder crackled across the Cape. Thrust built for four pounding seconds until Pad 14 released its bird, and in front of 50,000 watchers gathered at the Cape, Glenn surged into the sky.

JOHN GLENN'S REENTRY

JOHN GLENN REACHED orbit safely. After all the worrying, the danger was not the Atlas rocket; it was the capsule's heat shield that caused problems. The heat shield on the bottom of the capsule was Glenn's protection against the fiery trial of re-entering the Earth's atmosphere. As Glenn shot around the Earth in his Mercury capsule, the astronaut felt no buffeting or turbulence because he was above the air, flying through empty vacuum. The transition between realms was the tricky part. Moving some 17,500 miles per hour, the rocket didn't have engines powerful enough to slow it to a gentle stop so it could simply parachute back to Earth. That would have taken reverse power equivalent to what had gotten him into space, far too great a burden to carry into orbit. Glenn's onboard engines gave him only enough braking power to cut his speed by about 350 mph, lowering his orbit sufficiently to drop him into the tenuous air of the upper atmosphere. The problem was that when he hit the atmosphere, he would still be going considerably faster than a rifle bullet.

Air has deadly power at high speeds. The same soft breezes that lift airplane wings from the ground can slice like a knife at high velocities or take on the violence of an industrial sandblaster. "Demons waiting at the speed of sound" threatened to tear apart aircraft with a hail of blows. After Chuck Yeager cracked the sound barrier in 1947, Americans began cautiously exploring the supersonic realm, where planes move faster than sound itself. We began to measure speeds in Mach numbers, Mach 1 at sea level equaling 712 mph, or the speed of sound.

John Glenn hit the atmosphere at almost Mach 24. Air at this velocity does not merely tear off things like protruding antennas or strip the paint off the hull...it can carve the metal itself, turn it white-hot to molten liquid, eat its way through the ship and through the seat until you feel the back of your spacesuit melt, and you begin to be incinerated inch by inch.

Primitive man personified the forces of the natural world as gods and demons, dire powers that would exact retribution upon those who dared to defy them. How easily an ancient Greek would have seen vengeful forces in the annihilation threatened by reentry. The myths had already told the story of this launch in antiquity and of the price of challenging

The original Mercury capsule that carried Gus Grissom on a 16-minute suborbital flight in 1961 was recovered from a depth of three miles on the ocean floor in remarkably good condition. The capsule sank after a hatch cover was jettisoned prematurely and it filled with water. 1999.

the gods. Icarus dared to fly up into the realm of the sun on wings engineered with wax, then he felt his crude invention come to pieces in the intense heat above the clouds.

How strangely the myth now foreshadowed the adventure of astronaut John Glenn who flew in a capsule named for the ancient god of speed. A light on a console in Mercury mission control had come on, indicating that Glenn's heat shield had come loose. It told the launch controllers that the wax in John Glenn's wings was beginning to melt. Glenn had beheld the wonders of the heavens and had ridden the pathways of the stars; now he would pay the price and serve as a symbol and admonition to his people that they might learn humility.

The controllers didn't tell Glenn what was wrong, but when tracking stations kept casually asking him to check a certain switch setting, he became suspicious. The Mercury control engineers at the Cape got hold of Max Faget, the space capsule's chief designer, at Houston. Faget agreed with the operations team's plan: if the heat shield clamp had released prematurely, perhaps the shield might be kept in place during re-entry by the braking retro-rocket pack that was strapped over it. Instead of jettisoning the retro pack as planned, Glenn could just leave it on, and let it burn up during re-entry. The straps might help hold the heat shield on before they burned off.

Glenn followed the instructions NASA radioed up and plunged into re-entry. Strapped tight in his cramped capsule, he watched with alarm as glowing chunks of what he thought was his heat shield disintegrated and hurtled past his window. NASA waited spellbound as Glenn traversed the zone of isolation where re-entry's ionization of the air around the capsule created an impervious barrier to radio signals. Something came out the other side—it was Glenn and he was all right. To Glenn's profound relief the parachute finally deployed and he drifted down into the waters of the Atlantic, safe and sound. He had rewritten the myth of Icarus for the 20th century.

The United States erupted in pandemonium to cheer the hero who had defied the ancient gods and won. As ticker tape rained down by the ton, jubilant crowds crushed against the fenders of the cars in a New York City parade held in Glenn's honor. After all, here was a living symbol of what we could do. New York welcomed Glenn home in matchless style, but his Mercury-Atlas flight had set out for history from the one place that could make it possible. Cape Canaveral was, for the modern world, uniquely sacred ground…a proving ground not only for hardware, but also for the human spirit. The sky was no longer the limit.

Three more astronauts would follow Glenn into orbit on Mercury-Atlas launches, each mission lasting longer and stretching capabilities

←← **Astronaut John Glenn on his way to speak to a joint session of Congress is greeted by thousands despite the rain. Seated with him in the back of the car is his wife, Annie, and Vice-President Lyndon Johnson. 1962.**

Astronaut Gordon Cooper, carrying his portable air-conditioning system, leaves for Launch Pad 14. 1963.

in the new environment of space. Scott Carpenter, Wally Schirra, and Gordon Cooper had their adventures beyond the sky as the Cape made sure they got off the ground safely. After Cooper had flown the Mercury capsule to its limits with a flight lasting 34 hours, it was time for a new astronaut program, one with higher ambitions and a larger spacecraft.

GEMINI

7

PROJECT MERCURY HAD put the first Americans in space and demonstrated their basic abilities to survive and operate in that hostile environment. Shortly after Alan Shepard's momentous flight, President John F. Kennedy launched America on one of the greatest national quests of all time, committing the country to landing a man on the moon before 1970. NASA scrambled in the face of this dizzying challenge to deliver on this pledge. The moon-quest project took the inspiring name Apollo, but a tremendous gulf lay between the accomplishments of Mercury and a realistic chance of reaching the moon. To fill that gulf NASA initiated Project Gemini.

← Ed White, the first American to walk in space, was attached to the Gemini spacecraft by a 23-foot tether line and carried a Hand-Held Self-Maneuvering Unit to move about in space. 1965.

Project Gemini sought to achieve a list of very specific objectives, milestones that would build our abilities and technologies toward the goal of landing astronauts on the moon. Could we achieve rendezvous in space? Could astronauts work outside their spacecraft, floating in open space? Could crews live in space without serious medical consequences for the length of time it would take to fly to the moon and back? Could trained pilots maneuver their ships with precision and navigate between orbits without losing control? Could a spacecraft re-enter the atmosphere with sufficient accuracy to splash down within sight of the recovery forces?

Critical to Gemini was a next-generation spacecraft. The new Gemini ship was not just a two-seater version of the Mercury capsule, but a far more sophisticated vehicle. A Mercury capsule could be pointed in different directions with its small thrusters, but couldn't maneuver at all except to decelerate for reentry. The Gemini ship carried maneuvering thrusters, along with rendezvous radar and—astonishing for the time—an onboard computer. Operating the radar and computer took a second crewman, which is why the Gemini had to be a two-man ship. Powered by fuel cells instead of short-lived batteries, the later model Geminis would have long-duration capabilities. For an astronaut, the Gemini was a responsive precision flying machine, a real science-fiction-style spaceship.

All this capability and technology, together with the tanks and reservoirs and such, added up to weight—a lot of it. The Gemini capsule would get nowhere near space without a lot more rocket power than any variation of the Atlas could provide. If Kennedy's deadline was going to be met, there wasn't time to develop a dedicated Gemini rocket—but there was one on the shelf that would do.

The new missile program was called Titan, and the Martin Company developed the Titan I so quickly, with its first full-on launch in 1960, that it was almost ready by the time the Atlas was operational. The highly capable Martin operation then developed the even more advanced and powerful Titan II and had it flying in early 1962.

In the Titan II, an engine with twin thrust nozzles drove a large first stage; a second stage extended the missile's reach. Larger than any rocket ever built before, Titan II's second stage pioneered the technology of igniting a liquid-fueled stage at the extreme altitude of 40 miles. The system worked, and America's most potent ballistic missile easily had an intercontinental reach, with enough power to carry multiple warheads. Using room-temperature chemical propellants, it could also be launched literally within one minute of the order being given, while a defending Atlas had to be raised out of a silo, fueled up with cryogenic oxidizer and put through a lengthy countdown—during which it would likely be blasted by an incoming enemy nuke.

LC-19 AND THE LEANING TOWER OF GEMINI

FOUR NEW SITES arose on the Cape coast for this hefty new Cold Warrior: Pads 15, 16, 19 and 20. Each was a separate complex, rather than A/B pairs sharing a blockhouse. The launch control center stood 600 feet from the pad and took the form of an igloo fortress with 40-foot-thick walls, similar to the Atlas blockhouse. But the degree of fortification was still higher: a two-foot-thick solid steel door sealed the blockhouse with 20 tons of protection. Complexity also increased. The myriad connections to the rocket and its service systems required 800 miles of cable running between the blockhouse and the pad.

Martin's Canaveral Division built the Titan pads with a highly original form of service tower, one that pivoted on its base to lie flat against the ground when not in use. Previous service gantries rolled into place to lower their hinged work platforms around the rocket, and then simply rolled out of the way before launch. Martin's red-orange Titan tower instead swung through an impressive arc, rising high into the air to lift each rocket stage into place. The process looked oddly like an 11-story

Astronauts Virgil Grissom (foreground) and John Young board their Gemini 3 spacecraft in the white room atop the launch vehicle at Pad 19. 1965.

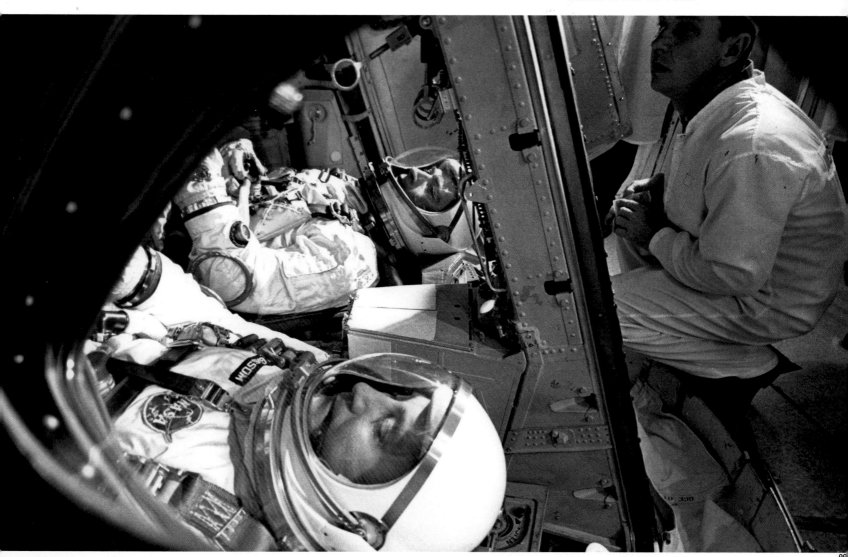

Gemini 12 sits on the launch pad as the service tower pivots down and out of the way of the blast. 1966.

office building sitting up from bed, like Florida's own "leaning tower of Gemini," a space-age New World answer to Pisa's monument in the Old World. The pivoting process was reversed shortly before launch, the service tower dropping down back from vertical to lie horizontal with only the separate umbilical tower remaining alongside the rocket.

With the Titan II, the Martin Company had produced a superb machine. This hefty piece of engineering could drive a multi-ton payload all the way into orbit, and, beginning in 1963, the air force made it the new backbone of America's strategic nuclear defense. The Titan II's muscle also got a job in the space program: modified for NASA, it would carry the astronauts of Project Gemini.

A LIGHTWEIGHT WHITE ROOM

MARTIN'S CANAVERAL DIVISION converted complex 19 into the Gemini pad, installing specialized equipment at many points to support the addition of the intricate space capsule to the top of the Titan II rocket. The blockhouse got new computers and systems monitors, while the leaning service tower would have to be extended to accommodate the revised rocket, which stood 10 feet taller with the spacecraft sitting on top. The service tower would also need a large white room enclosure to protect the space capsule during launch preparations. All this would involve increasing the height of the leaning tower by 28 feet. Building the Gemini white room at Pad 19 presented a special problem since the 150-horsepower winch that raised and lowered the pivoting service tower could carry only the existing tower's limited weight. Martin couldn't simply tack on a white room at the top of this tower, because when activated the overweight structure would have just stayed on the ground, snapping the straining winch cables or burning out the winch motor.

Therefore Martin Canaveral had to build a complete four-story white room, and add a five-ton crane and an elevator, without adding any weight to the tower. They accomplished this by chopping off the top 19 feet of the old steel tower and replacing the missing section with a new extended structure made almost entirely out of lightweight aluminum. The engineers were so strict about controlling the weight of the new structure that the weight of every rivet had to be accounted for before it went into the new white room construction. Lightweight solutions were adopted in every way possible. Even the elevator to the top of the gantry was made to run on one rail rather than two, a somewhat alarming compromise to those who had to use it frequently. In the end, the new white room weighed no more than the steel chopped off the old tower, and, when completed, the modified monument rose into the air just as easily as ever. The new structure provided a "roll-up garage door" to slide down over the space capsule once the tower was in place, sealing it in for climate control. While the whole system worked, pad crews had to be certain they didn't leave anything onboard the gantry, since with the 90-degree tilt of the tower, anything left behind would probably slide out and disappear when the tower dropped. It was the last time this pivoting tower design would ever be used.

Project Mercury had involved the risks of the deep unknown; Atlas had brought the risks of experimental technology. Gemini would continue to face risks, challenges and the unknown, but growing sophistication and increasing experience gave NASA a new level of mastery over its tools.

NASA would borrow the Titan II from the air force, but the missile would undergo a stringent metamorphosis in order to earn "man-rated" status. A new malfunction detection system was installed to improve safety. Martin took great pains to incorporate redundant components and to ensure reliability to the greatest possible degree. This work made the Gemini-Titan II an improved version of the rocket design, a worthy chariot for the space warriors brave enough to ride it.

When it came time to launch the first piloted Gemini spacecraft, NASA chose Mercury veteran Gus Grissom as its commander. His predecessor into space, Alan Shepard, was off flight status due to an inner ear equilibrium problem, and his famous successor John Glenn was considered by the White House too valuable as a national hero to risk on another flight. That left Grissom, the taciturn Indiana astronaut, as the senior member of the team, the man who would prove the new capsule. Two unmanned ships were sent up on test flights before the craft was ready for its first astronauts.

Both of the Gemini-Titan II's rocket stages were test fired on the ground at Pad 19 in advance of the flight. The first stage fired its twin-nozzled engine over the main flame deflector in the same position it would occupy during an actual launch. Alongside the main leaning tower, a shorter copy lifted the smaller second stage into place over a second flame deflector where it, too, proved its worth by firing through a rehearsal sequence. Only after this acid test was the second stage lowered back to horizontal, removed from the shorty gantry tower, and inserted in the lowered main tower to be swung back up to vertical again and joined with the first stage.

Grissom's spacecraft was the last element to be added to the stack. When all was readied, the spacecraft arrived via truck at the foot of the pad. A pad deck crane hoisted it up to pad level and into the reach of the gantry tower's white room crane, which then reeled it ten stories up. At the top of the tower, the crane carefully drew the capsule into the white room where it would be eased down and bolted onto the rocket.

On March 23, 1965, the water deluge system flooded the blast zone to reduce flame damage to the pad, and the Titan II rocket of Gemini 3 flared into powerful life. Orange smoke and cottony clouds of steam from the seared deluge water billowed from the base of Pad 19, and then Grissom's 108-foot space vehicle lifted off with startling effect. The Titan rocket used hypergolic fuel, chemicals so reactive with each other that they burst into fire on contact without an igniter. These hypergolics burned with a largely transparent flame, making Grissom's rocket appear to rise on no more than slender flares at its base. It left no trail as it elevated into the sky like a helicopter or some extraordinary new antigravity device.

← Gemini 6 lifts off, carrying astronauts Walter Schirra and Thomas Stafford, who will rendezvous their spacecraft with the Gemini 7 spacecraft already in orbit. 1965.

Media at the time generally framed Gemini as a NASA operation, but in truth it was a multilateral team effort, a NASA spacecraft and crew launched on an air force rocket from an air force pad by a contractor launch team. That launch team was the Martin Canaveral Division. The Martin Company took on the Gemini assignment with the greatest commitment to performance and safety. It developed incentive programs to encourage its manufacturing force to do work with "zero defects," and it involved the astronauts in presenting awards, helping workers appreciate the importance of their work and the human life that depended on their workmanship. Martin considered the workforce to be as important as the components they were producing and undertook to train and rigorously test everyone on the rocket-building team. Anyone who could not meet the required standards was removed from the program. The result was a set of excellent, dependable Titan II rockets throughout the Gemini Program, and the Martin Canaveral Division likewise delivered a superb record of launch performances. Many of the countdowns went flawlessly, an exceptional performance record in these early experimental days of rocketry.

An employee of McDonnell, the company that built the Gemini capsule, looked after the spacecraft while it was on top of Martin's rocket. Guenter Wendt was a no-nonsense, ex-Luftwaffe mechanic with a penchant for strict organization and thorough planning. An uncompromising individual, he also represented the best of the "team player" spirit that would take NASA to the moon. Wendt, a pad leader, was in charge of the white room, the spacecraft level of the gantry. The wise-cracking astronauts took to calling him "der Pad Fuehrer" in honor of his strictness. Wendt was good-natured and jovial with the astronauts, sending each one off with a joke or present when he led the team that sealed up their capsules on the day of liftoff, but he ran his section of the pad with absolute discipline. To Wendt, politics were meaningless in the face of the job's demands; he knew that lives depended on strict procedures, especially up at his level of the pad in close quarters with the spacecraft. He brooked nothing out of line. When arrogant officials visiting from Houston got in the way of his work, he had them escorted right off his pad by Cape security. "Don't touch the spacecraft, don't lean into the hatch opening and make sure your pockets are empty," were his basic rules for visitors. He found that almost everyone but "junior bureacrats" could follow them just fine.

Despite the prolonged delays that had slowed the starts of the program, NASA hoped to fly Gemini missions on a bimonthly basis, an absolutely unprecedented pace for manned missions. Indeed, the program would end up sending "twenty men in twenty months" into space, reflecting the Cold War sense of urgency. This ambitious schedule would place great demands on all involved.

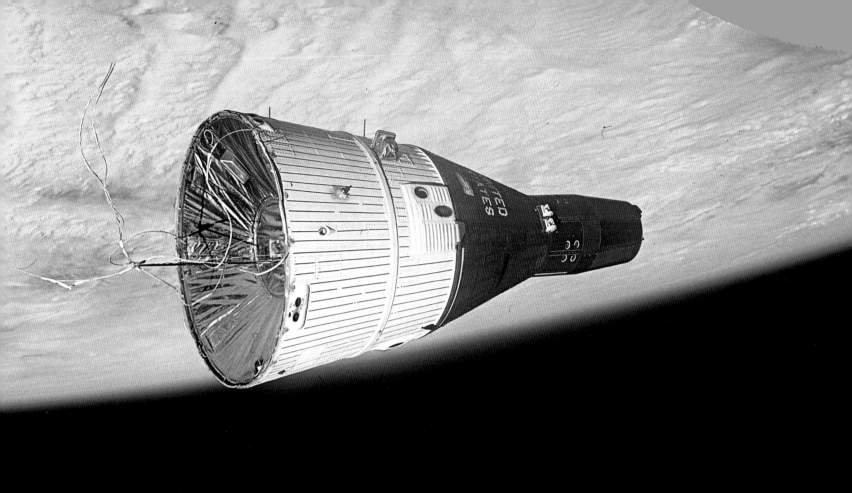

This photograph, taken by Frank Lovell and Frank Borman in Gemini 6, shows Gemini 7 in orbit 160 miles above Earth during the first successful rendezvous mission between two spacecraft. 1965.

Ed White, pilot of Gemini 4, floats in the zero gravity of space. 1965.

NINE-DAY TURNAROUND

PAD CREWS WORKED hard before and after a rocket launch. Preparing a rocket for its heroics in flight and in space involved systematically elevating the levels of danger involved with a system that started out as a set of empty tanks on arrival and peaked at the moment of launch as a live bomb sitting on the pad with the power to blow the complex into scrap. The work began at Pad 19 with installing each of the two Titan stages in its own leaning tower and test firing them through trial runs. The two stages were combined in the main tower along with the spacecraft and carefully locked together into a single unit. The pad crew then fitted the Gemini-Titan II combination with its extensive set of explosive bolts and pyrotechnics to cut the stages apart during staging and to provide the booster with self-destruct capability. For safety, all this had to be done at the pad — no earlier — and had to be carried out with considerable care. Finally, the pad and blockhouse teams ran the vehicle through a long checkout to ensure that the entire rocket and all its systems were A-OK.

After launch, the scorched pad looked like a disaster area. While the leaning service tower could recline out of the way before launch, the umbilical tower that carried the rocket's fuel connections had to endure the blast and stand fast during launch. In spite of the firehoses installed up and down the launch umbilical tower to drench it during the fireworks, thousands of pounds of cables were fried, burnt and ruined each launch. It was just part of the price of sending a rocket up. Between the rocket installation and the cleanup after a launch, turnaround at the Gemini pad normally took about three weeks.

After Gus Grissom's short proving flight, Ed White performed the first American spacewalk during Gemini 4, following hard on the heels of the Soviet spacewalk in mid-March. The U.S. was still behind, but it was closing the gap. Gordon Cooper's Gemini 5 stayed in orbit for almost eight days, testing the maneuvers that would be used for rendezvous.

Rendezvous in space was a prime goal of the Gemini program to prove that the rendezvous necessary for a moon mission could be accomplished. Rendezvous requires a target, and the standard Gemini target would be an unmanned stage called Agena that had been carried into orbit by an Atlas rocket. Convair would launch them from Pad 14, the same complex that had sent up John Glenn and the rest of the Mercury-Atlas missions, and which was now converted to support the Atlas-Agena combination. The Agenas would be sent up in advance, and the Gemini crews would fly up and hunt them down, if they could.

Wally Schirra's Gemini 6 was to be the first rendezvous mission, with

A profile view of the Agena Docking Target Vehicle as seen from the Gemini 8 spacecraft during rendezvous in space.

Tom Stafford as the pilot. On August 25, 1965, the Agena target blasted off on an Atlas from Pad 14, just 90 minutes before Schirra's ship was scheduled to blast off after it from Pad 19. The Agena crackled its way up into the sky, and exploded into a million pieces.

Without a target, Gemini 6 was useless, but NASA's contractors had a bold idea. Instead of the lost Agena, Gemini 6 could chase Gemini 7. Martin, confident in view of its superb operations record, had already suggested the idea of a salvo launch, and McDonnell insisted to NASA that the time was perfect. Despite the Agena explosion, a double launch would accomplish both missions without an overall schedule delay. As the builders of the Gemini spacecraft, McDonnell's engineers were in a position to appreciate the possibility now before them. Gemini 7 was one of their advanced model spacecraft, carrying fuel cells for extended capability, and it was intended to fly a 14-day endurance mission. That period would give the pad crews enough time to refurbish the pad and send Gemini 6 up to chase Gemini 7 down and accomplish the rendezvous. To allow some margin for error, they would have to turn the pad around in nine days.

Guenter Wendt took one look at the schedule and said, "Oh, man. You are crazy." If nothing—absolutely nothing—went wrong or required major repair, it was just barely, possible in theory. These were daring days at the Cape, and NASA went for the plan. The pad crews took Schirra's Gemini 6 rocket down and stored it next door at Pad 20, stashing it in a hangar while they installed Frank Borman and Jim Lovell's duration-flyer Gemini 7 and its rocket on Pad 19. After the blastoff, Wendt's crew worked desperately against the clock, beating their crazy deadline by an entire day. The pad team's accomplishment has passed into Cape legend, a marvel and milestone of the great days of the Space Race. Martin hauled Gemini 6 back over from Pad 20 and everyone set her up again.

Rather than sporting an escape tower, the Gemini capsule featured ejection seats. The Titan II hypergolic propellants would not burn as explosively as Redstone kerosene, so there would be time for an ejection seat to clear the blast. The launch escape tower was complex and difficult to work with and required a long checkout, so the capsule designers were glad to be able to do without it. The ejection seats would work fine during the earliest stages of flight, before velocity got too high, but they presented problems for a pad bailout, and the pad was one of the most dangerous places for a rocket—and for an astronaut. Wally Schirra had this well in mind when he boarded his ship on December 12, 1965. As one of the final pre-launch procedures, before closing and sealing the spacecraft hatches, pad leader Guenter Wendt and his crew of pad techicians removed the seven pins that kept each ejection seat inert. The seats and their rocket systems were now live. Schirra knew that in the case of a pad disaster he would pull a D-ring below his seat. Explosive bolts would blow the hatches off, and the seat rockets would fire to blast both him and his crewmate Tom Stafford clear of the conflagration. NASA had set aside part of the pad complex as the "pilot bailout area," although they expected the ejection seats to carry the astronauts all the way out to an ocean landing.

But there were problems with this concept for a pad escape. First, at this low altitude, bailout with an ejection seat risked not having enough time for the chute to deploy if the seat rockets didn't carry you high enough. A parachute doesn't do much good if it doesn't deploy before hitting the ground. The ejection seat system had not actually been man-tested, and it promised a rough ride at best, with possible injuries even in a survival result. In one test with a dummy, the hatch had failed to jettison and the seat had punched its way through it. Another factor to consider was that the dummy tests had been made with the capsule full of inert nitrogen. During launch the pilots were surrounded by pure oxygen, and there was a chance that if they fired, the seats would emerge as flaming candles. Deciding to pull the D-ring meant choosing between certain death while staying in place or possible death when ejecting. A further problem with the seats was that ejection would destroy the capsule. You don't set off twin rockets and a round of explosive bolts inside a cramped spaceship cockpit without doing damage; a bailout would pretty well wipe out the capsule, leaving Borman and Lovell to float around in space with no one to come chase them. This would ruin the rendezvous mission Schirra had been assigned.

Schirra got a chance to make a critical decision about ejection when launch came—"3…2…1…liftoff!"—and his rocket shut down. The mission clock running on his dashboard indicated that the tail plug had been pulled out of the rocket by liftoff movement, meaning that they were now in the air and about to fall back, with the Titan II rocket's 150 tons of propellant about to burst into a horrible fireball. The rules said eject immediately. But Schirra did not pull the D-ring. One instrument was telling him there had been no liftoff. He later said, "I recalled from my Mercury mission the sensation at liftoff, and it hadn't occurred." He wasn't going to pull that D-ring until he knew for sure they were in trouble. In every NASA simulation of such a situation before and since, the pilot followed the rule book and ejected, but Schirra listened to his experience and he didn't pull the ring. The rocket hadn't lifted off—faulty indicators had lied.

"How are you guys?!" asked launch control.

"We're just sitting here breathing," Schirra replied. NASA never saw a cooler hand in the midst of danger. This was the kind of situation that proved the value of a superb astronaut. The launch would have to be recycled, but the capsule was in perfect condition—and the crew was little the worse for the adrenaline rush. Schirra's cool hand saved the mission. After this false start cost the three-day margin of error, Schirra was blasting into orbit with no time to spare.

SPACE RENDEZVOUS INVOLVES intricate orbital calculations, precision maneuvering and a delicate touch at the controls. The basic physical law governing orbital navigation is that lower orbits are faster, and higher orbits are slower. Counter to intuition, you can't simply increase your speed to catch up with your target; instead of going faster, the increased momentum will translate to a higher orbit at slower velocity. You'll see your target slip away ahead of you until it drops past the planet's horizon, and circles the earth to come up behind and below you. To go forward and catch up, you have to put on the brakes, which drops your orbital altitude...and speeds you up. During missions, an onboard computer helped calculate these maneuvers.

Wally Schirra knew how difficult space rendezvous would be, and he and Tom Stafford had practiced in the simulator over and over and over again. All the astronauts brought their different personalities and strengths to the work. Schirra's aptitude was focused precision flying, which he achieved through dedicated practice and absolute mastery of the mission and machine mechanics. Science and exploration could wait for the other guys. Schirra was a pilot's pilot. He was going to make this rendezvous work and pave another stretch of the road to the moon.

Each Gemini mission took America higher and farther. The whole program was designed to gain specific capabilities NASA needed before it sent up the Apollo moonships. Every mission had the express purpose of developing or advancing one or more Apollo capabilities. We learned to maneuver in space, we learned to walk in space, and, most importantly, we learned to rendezvous. Schirra's cool work and light touch on the thruster controls made it look easy, but the rendezvous was nothing short of virtuoso flying. By the fourth orbit, right on schedule, he was nosing up to within 12 feet of Frank Borman's Gemini 7 capsule.

"How's the visibility?" asked Borman.

"Pretty bad." Schirra radioed back. "I see through the windows and see you guys inside!"

Suddenly the rendezvous component, which had made Apollo's route to the moon seem so difficult, was possible. It was a wonderful milestone. Later missions saw astronauts practice docking with target stages—Neil Armstrong was the first to carry out such a dock. Astronauts also learned to handle spacewalking, which was so difficult at first that it threatened Gene Cernan's survival as he struggled with a fogged faceplate. Using handholds, footholds and new movement techniques developed by Buzz Aldrin, astronauts on later space missions found that suddenly spacewalking had become easy too. Gemini took unknowns and systematically conquered them. Step by step, it put Americans closer to the moon, blazing the trail for Apollo. Every one of its missions sailed skyward from Pad 19.

With the conclusion of Gemini, NASA found itself on the threshold of Apollo, the program that would finally take man to the moon. Again, the focus of activities moved northward, to a new pad for a new and larger machine... at a site that would exact a greater price from all involved.

→ Elation is written all over the faces of astronauts Frank Borman and James Lovell aboard the aircraft carrier *Wasp* after completion of their 14-day mission in space. 1965.

JOURNEY TO THE MOON

The Apollo program would leave its mark on the Cape more powerfully than anything else before or since. The facilities built to receive, prepare and launch the Saturn V were so large they could be seen from orbit.

8

SATURN

THE ATLAS AND Titan II rockets had carried U.S. astronauts far beyond the suborbital hop made by Wernher von Braun's Reliable Redstone missile back in 1961. After the early dazzling performances of America's first satellite launch and its first men in space, von Braun and his team had been busy at Huntsville's new NASA Marshall Space Flight Center, crafting the prototype of a rocket that would carry the first Apollo capsule—a three-man spacecraft much heavier than even the Titan II could lift. Von Braun's new rocket was called the Saturn, and when it first appeared at the Cape in the fall of 1961 even the seasoned veterans were impressed.

← Apollo 16 lifts off for its 12-day mission to the moon. 1972.

←← The Apollo 4 Saturn V rollout proceeds to Launch Complex 38A. 1967.

VON BRAUN'S SATURN ROCKETS

THE SATURNS WERE envisioned as a family of launchers and in practice three major types would be built: Saturn I would test the concept, Saturn IB would take the first Apollo capsule aloft, and the Saturn V would send men to the moon. The three each got their numbers from concept notations, so there were never any Saturn II, III, or IV rockets built. A rocket larger than any Saturn, called the NOVA, was originally envisioned as the moon rocket, but by 1962 it was apparent that the NOVA could not be prepared by Kennedy's deadline – it was also realized that a smaller ship than the giant lander first considered could be sent to the moon. The Saturn type C-5 would be powerful enough to carry the smaller lander, and it could be developed within the deadline, so the renamed Saturn V became the moon rocket.

The Space Race was all about building maximum rocket power as quickly as possible. Von Braun had devised a shortcut to leapfrogging the Titan II, creating a rocket powerful enough to test the Apollo spacecraft in orbit ensuring it would be ready in time for the arrival of the full Saturn V moon rocket. Von Braun knew that developing a new rocket propulsion system cost critical time, time that they were already spending on the development of the Saturn V. Every advance in rocketry was risky and expensive, so it was always an advantage to work with known entities. To create a practical interim Saturn within the tight deadline, von Braun proposed to more or less bundle a bunch of Redstone missiles together and call it a first stage. The real Saturn I was not quite as simplistic as that sounds, but it really was built out of a set of eight stretched Redstone tanks strapped together, each with its own new engine. The core was a stretched Jupiter tank.

The eight new engines were fresh off the workbenches of Rocketdyne. Rocketdyne dramatically surpassed the old "cast-iron" V-2 engine with an intricate design in which the engine bell was built out of fine circulating propellant tubes to improve performance. A superior fuel injection and ignition system achieved better combustion too, putting Rocketdyne on the road to becoming a byword for the world's best rocket engines. American engineering had risen to the task with admirable force.

The Saturn I would present a cluster of eight of these Rocketdyne engines, although critics said there was no way all eight would work together smoothly. Von Braun countered that every engine had seven backups, therefore, a failure of any one would not be catastrophic to the mission. Nonetheless, the bundle of rockets was in truth a somewhat improvised solution – and it looked it. It would work for the short term,

Five pioneers of America's space program pose with models of the rockets they created. From left to right: Major General Holger Toftoy (standing), Dr. Ernst Stuhlinger, Professor Herman Oberth, Dr. Wernher von Braun and Dr. Robert Lusser. 1956.

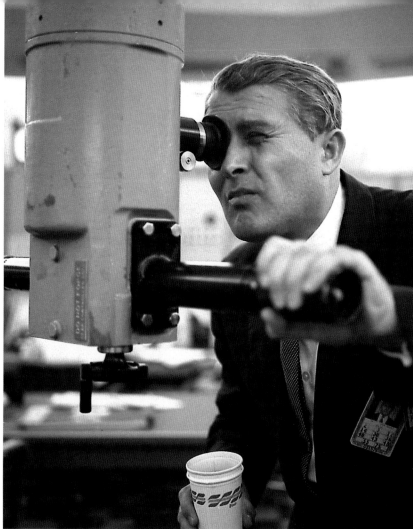

but once more powerful engines were developed, NASA would never return to such a method again.

Von Braun's new Apollo Saturn rocket would need a new launch complex, and when the time came, it was laid out at the Cape and named Pad 34. It had a domed blockhouse like all the ICBM pads, and von Braun had a periscope through which he would watch his rocket like a submarine commander. Conservative German engineering made Launch Complex 34 at heart a giant version of the same launch setup that von Braun had been using ever since Peenemünde. The four-legged square launch stand had grown four stories tall, but it was still the same basic stand-in design. It still had a flame deflector underneath, only instead of a small metal pyramid there lay a gigantic two-sided inverted V. Why two-sided instead of four? Because the new rocket would pack a tremendous blast, and if the flame deflector shot the blast out in all four directions, it would spoil von Braun's view from the blockhouse periscope! The V was thus angled to shoot the blast out only to the sides so he could watch.

The Saturns were so powerful that if one of them exploded, it would wipe out the entire complex. The traditional "nearby backup pad"

wouldn't provide complete insurance with the potential destructive power of a Saturn. One such explosion at LC-34 could delay the program for a year—a critical loss in the moon race. Accordingly, a backup Saturn launch site was approved: Launch Complex 37.

The place where they wanted to put Pad 37 was a swamp. To stabilize the foundations of the site, the engineers packed the sand down with a process called vibroflotation, which compacts the sand until it is crushed together like concrete and can bear heavy loads. To build LC-37 they had to move 220,000 cubic yards of sand. There were 2,000 tons of steel in the foundations alone, and the 10-million-pound service tower stood 328 feet tall. As far as anyone knew, it was the biggest thing on wheels in the world. This is what it would take to service von Braun's Saturn I and IB rockets…and they were only junior heralds of the coming Saturn V. As with so many other impossible tasks with Apollo, the U.S. Army Corps of Engineers faced new challenges and built equipment of magnitudes that had never been built before—and did it on schedule. In 1963, they brought in LC-37 on time.

Pad 34, however, would remain the primary Saturn I launch site. It was here that the Apollo crews trained for the first missions in the new, super-complex, three-man Apollo space capsule, and it was here that we would be reminded that an astronaut is at risk as soon as he boards the spacecraft.

Guenter Wendt had not been invited to perform his role as pad leader for the spacecraft levels of the gantry at Pad 34. There was a new contractor involved, and North American Aviation had new ideas about how to do everything. Consequently, Wendt was not looking after the white room and the tower safety systems when Gus Grissom and his crewmates, Ed White and Roger Chaffee, boarded their Apollo capsule atop an unfueled Saturn IB to carry out another full-dress rehearsal of the launch process on January 27, 1967. The new system was incredibly intricate, and although it was due to be launched in February much of it was not yet working properly. Communications were poor, and Grissom was frustrated with the inadequate technology. It was another long day of hard work, but it was just another stretch to get through…until the fire started. The capsule was filled with pure oxygen for a pressure test, and suddenly the cramped interior became a raging inferno.

The three astronauts were bolted into the capsule, and there was no way of getting them out quickly. The gantry crew struggled with the hatch and tried to assist, but when the cabin ruptured and spewed dense toxic smoke they were driven back. It was over five minutes before they could get it open, and by that time it was far too late for the men inside.

The engineers and the contractor North American had made mistakes

← Apollo astronauts Gus Grissom, Ed White and Roger Chaffee during tests of the Apollo 1 command module. 1967.

Apollo 1 command module after the fire that killed astronauts Gus Grisson, Ed White and Roger Chaffee during a routine training session.

with the design of the Apollo capsule. The analysis afterward suggested that wiring with damaged insulation had caused an arc beneath Gus Grissom's couch and, in the pure oxygen environment, that was all it took to touch off a "flash fire" holocaust. To everyone's horror and astonishment, the first Apollo crew had been lost on the ground. All launch pads were baptized with flame. For its first manned mission, Pad 34 was inaugurated with a pyre. The Cape was cast under a pall; all felt the weight of what had happened.

In the wake of the Apollo 1 fire, the Apollo capsule was thoroughly redesigned, until it became the safest spacecraft America has ever flown—far safer than the space shuttle that would follow it. In particular, the old hatch was replaced with a fast-action outward-opening one. Wally Schirra would command the first manned mission after the Apollo 1 fire, and he personally made sure that his ship was built properly, spending many days at the North American Aviation factory in California where it was being constructed.

At the insistence of the astronauts, Guenter Wendt was hired by North American to supervise the spacecraft levels of the Pad 34 gantry tower. They knew Wendt would not tolerate inadequate disaster planning, and would do whatever it took to ensure safety whether his company superiors thought of it or not. For years to come, Wendt would be a thorn in the side of any bureaucrats who placed policy or privilege above discipline and safety, but he stood as a living symbol of the spirit it would take to get us to the moon.

Wendt was there to bid *au revoir* when Wally Schirra boarded his redesigned Apollo capsule. Schirra knew exactly what he wanted to christen his ship. Although NASA had forbidden it, and it would always be listed as Apollo 7 in the books, it was the *Phoenix* to its commander. It would rise quite literally from the site of the ashes of Apollo 1 and his astronaut colleagues. When Apollo 7 lifted into the skies in October 1968, there could surely be no clearer statement of NASA's undaunted determination.

Pad 34 was deactivated in 1968 after only seven launches. The complex had seen NASA's greatest failure, as well as its greatest spiritual success. There were still a few Saturn IB rockets in storage—von Braun had wisely built backups. In spite of their unprecedented size, his Saturns never failed, so all the spare rockets were still available. NASA would put these to good use in good time, but no more would ever fly from Pad 34. The site was retired and, over the years, the corroding metal structures were removed and scrapped. Today all that remains is the skeleton of the complex: the igloo blockhouse, the wide concrete apron, the great flame deflectors and the giant concrete version of the Redstone launch pad. It looms out in its lonely spot like a modern Stonehenge, a cryptic symbol, the meaning of which can only be read by those who know the history. It serves as a monument, a fitting defiance of time to mark the place that burned with the flames of both the pyre and the phoenix.

Dr. von Braun
and President
Kennedy during the
President's tour of
Complex 37. 1963.

PREPARING APOLLO
FOR THE MOON

THE CAPE HAD seen a multitude of designs before the advent of the Saturn V, but the moon rocket stood apart from the others the way George Washington stood above the captains of the Revolutionary War. Like the father of this country, the moon rocket ranked alone. The Saturn V of the Apollo moon landing program would leave its mark on the Cape more powerfully than anything else before or since. The facilities built to receive, prepare and launch this leviathan were so large they could be seen from orbit. Even though the equipment has been modified for new purposes following the conclusion of the Apollo project, the Apollo Saturn V launch complex forms the core of Kennedy Space Center today. The moon rocket was gigantic in every respect, and its magnitude forced a rethinking of many operations that had become standard at the pads of the Cape.

Apollo planners had to devise a whole new launch concept to fit this extraordinary instrument, and they had to do it without delay, for

← This view of the entire rocket shows the 36-story Apollo/Saturn 501 in High Bay No. 1 in the VAB, in preparation for rollout.

Workers maneuver the giant sprocket and attached gear into position to turn one of the eight belts on the Crawler-Transporter. 2004.

Kennedy's deadline for the moon landing was ticking away with every passing day. The sand was falling irrevocably through the hourglass, and the miracle had to be delivered on time. NASA and its contractor allies would not disappoint. Inspired by the sheer daring of the entire enterprise and by the faith of the President who had told the world he believed they could accomplish the impossible, Apollo's designers and creators infused every part of the Saturn V operation with bold spirit, vision for the future and dedication to building works of surpassing excellence.

CRAWLER-TRANSPORTER

While the Mercury flights were ascending into space on Atlas intercontinental ballistic missiles, Apollo leaders were actively working on designs for the unprecedented moon launch complex. One of the most critical elements of the Apollo launch system would be the means of transporting a fully assembled moon rocket from the assembly building out to the pad. Kurt Debus wanted the matter settled, and settled quickly; he knew that the transporter would be a key issue and, indeed, one of the most technically challenging aspects of Moon launch engineering.

← The Crawler-Transporter fully loaded with the Shuttle and Mobile Launch Platform. 2005.

↓ The gigantic Crawler-Transporter tries out a new set of shoes by carrying the unloaded 8 million-pound Mobile Launch Platform along the crawlerway. Each shoe weights 2,260 pounds and the crawler has 456 of them. 2005.

The weight of the Saturn rocket was going to be astronomical, even without its fuel. Together with its launch tower and launch platform, the total would run to 12 million pounds. Under the weight of a Saturn V, trailers that hauled earlier rockets would be squashed flat, their rubber tires instantly popped like so many balloons. Rail systems had been used to move launch service towers ever since Peenemünde, but the analysis for an Apollo rail system did not look good. The engineers could too easily imagine the rails being squashed into the unstable ground by the weight. The natural alternative seemed to be a barge system, since water has been used to bear the greatest weights ever since the Egyptians shipped obelisks on the Nile. However, when the barge studies got down to figuring out the implications, it started looking like a mess too, with maneuvering difficulties and severe problems managing the launch blast in the canal.

From out of the coal fields of Kentucky came reports of a gargantuan device that crawled on tank-like treads, a behemoth so large it ate landscapes. Manufacturer Bucyrus-Erie had heard about NASA's dilemma with the transporter and invited Debus' people out to take a look. Not only were the stories true, but the mining company had an even bigger monster under construction nearby, and the new one was being built to handle a load even greater than that of a moon rocket. Debus' team were delighted to find engineers who were actually dealing with forces of Apollo magnitude and building to match them. An Apollo crawler would be state-of-the-art without being a dangerous experiment like the megabarge. Debus' team later realized that they could build the crawlers separate from the launch platform, and thus build only two of the transporters as ferries rather than three self-propelled launch platforms. When the numbers were run, the crawler was the clear winner.

Two identical machines were constructed in Ohio by competing bidder Marion Power Shovel and then broken down for shipment to KSC. Each crawler was a large, flat squarish shape, 131 feet long, battleship gray, with a load-bearing deck on its back the size of a baseball diamond. Walkways ran all the way around the sides of the machine, and it sported cabs on two opposite corners for driving either forward or backward—it didn't do turnarounds. It stood off the ground on four "trucks," which were double sets of treads at each corner of the square. Each tread unit was 10 feet tall and 40 feet long. One had to climb a retractable gangway to board this monstrosity, and it was like watching a factory come to life to see it begin to move. This was a carrier vehicle worthy of a Saturn V.

THE VEHICLE ASSEMBLY BUILDING

The Redstones had been assembled in the Cape's alphabet hangars. Larger rockets like the Titan II and Saturn IB wouldn't fit in such hangars, so these vehicles were put together by service tower cranes out on their pads. Checkout times were stretching into weeks and months with these increasingly complex systems, and this left the rockets sitting out in the salty, corrosive Florida weather for dangerously long periods. Therefore, for the Saturn V, Kurt Debus devised a new concept: the Moon rocket would be assembled under protected conditions in an assembly building that could contain the entire rocket in a vertical position. The whole space vehicle would then be moved as a single unit to the launch site, keeping exposure time to a minimum. The "mobile launch concept" would also keep the Saturn V pads from being tied up for too long by each launch operation. Several rockets could be assembled concurrently in the new assembly building, and a continuous series could be launched with just two pads. NASA planners were looking ahead to potentially dozens of Saturn V launches each year for an expanding space program. The assembly building, like all the Apollo facilities, was designed and built with future expansion firmly in mind. The novel concept of preparing the rocket standing up won the Apollo hangar the name Vertical Assembly Building. Later it was changed to "Vehicle" Assembly Building for reasons no one can remember. Either way, the place would be better known to the KSC crews as simply the VAB.

Two 250-ton bridge cranes strong enough to lift the gigantic stages of the Saturn V would be mounted in the roof. Work platforms would slide into place and close tightly around the rocket, giving engineers 360-degree access to five levels of the vast system for checkout. When the three stages of the Saturn V were assembled, the moon lander and the Apollo mothership would be stacked on the very top of the rocket, locked into place and topped by an escape tower. Once complete and fully checked out, the assembly could be taken to its launch pad.

This all sounded straightforward enough until the scale of the VAB sank in. To hold the 36-story spire of the Saturn V together with its even taller umbilical tower—and give the cranes room above it all—the building would have to tower 52 stories high. This would be for some time the largest building in the world by volume, enclosing over 129 million cubic feet of space, its 8-acre footprint larger than two football fields.

The designers gave the building not one or two, but *four* bays inside, in keeping with the anticipation of a long march of Saturn Vs serving the nation's future heavy-lift space needs. Six bays were considered, but four slots were approved as sufficient for the time being. The building was designed and oriented on its site so that it could be easily expanded to six,

eight, or more bays as needed.

Three of the VAB bays were outfitted during Apollo, while the fourth was reserved for any new versions of the Saturn V, which could take slightly different configurations and need new shapes of work platforms. Looking ahead, as always, von Braun was thinking of beefing up his moon rocket into an even more powerful Mars rocket. In addition to the reserve bays, von Braun made sure that the VAB roof could be raised for the even taller rocketships to come. The building's approximately 98,000 tons of steel included a full set of girders protruding from the flat roof, ready for upward expansion. They can still be seen today, pointing the way to the future.

The four internal bays, arranged as a square and enclosed, produced a 525-foot-tall box shape that was 518 feet square at its base. For the engineers, the building's broad faces were the equivalent of the biggest sails in the world, with more than one million square feet of area. Like a colossal clipper ship, the VAB was going to catch the winds with a vengeance, including the hurricanes that helped make Florida such an exciting place to live. To withstand hurricanes, the VAB would not only need to be sturdy, it would need better anchorage than ordinary foundations in the sandy muck that lay at the construction site. Undaunted, the builders drove 4,225 steel pipe pilings all the way down through the sand and mud to the bedrock shelf that lay 150 to 170 feet below the surface. The anchorage pilings were 16 inches in diameter, and the builders kept driving them until they had embedded a cumulative length of 123 miles' worth. Anchored to solid rock, the VAB structure could withstand sustained 125 mph winds, equivalent to a hurricane that could wipe out an ordinary city.

The four main bays of the VAB constituted the high bay area. A 210-foot-tall low bay annex, which formed the arrivals dock where rocket stages would be brought in from outside, led into the midst of the bays via a transfer aisle. The low bay held eight checkout cells for stage preparation. Test-fit equipment in these cells simulated the stage interfaces each stage would lock into and certified that it would fit properly into place once assembled. Only when checked and accepted did a stage pass on to the high bays for assembly.

Earlier rockets had been stacked vertically at the pads, where the standard procedure ever since the Navaho winged missile of the 1950s was to bolt or clamp them to the launch stand, and keep them locked in place for a few seconds after ignition. Once thrust built to full, the bolts were cut by explosive charges or the clamps opened to release the rocket. This technique saved many rockets from faltering at launch during the crucial first moments of thrust buildup and stabilization, and would be employed for

Dr. Debus, Kennedy Space Center's first director, adds his name to the thousands of signatures on a 38-foot steel beam used to mark the completion of the VAB. 1965.

← The VAB with the Launch Control Center and Service Towers as seen from across the Turning Basin during construction. 1965.

The Apollo/Saturn V Center, one of the latest additions to the KSC Visitor Complex, contains a complete 363-foot-long Saturn V rocket among the displays in its 100,000 square feet of space. 2005.

the Apollo rocket as well. Stacking the Saturn V in the assembly building required a portable stand on which to mount and transport it, and this took the form of a huge gray rectangular box called the Mobile Launch Platform. This 25-foot-thick launch platform, a major piece of construction in its own right, featured a 45-foot square hole cut through its center where the rocket blast would pass during launch. Around the rim of the blast hole were the clamping points, and it was here that the first stage of the Saturn V was locked into place to begin a moon rocket stack. Apollo 11 was stacked on Mobile Launch Platform 1 in High Bay 1.

Hoisted up, moved into position, lowered into place and carefully secured, the three stages of Apollo 11 were stacked one by one until the body of the Saturn V took shape. Alongside it in the universe of steel latticework that was the VAB stood the even taller Launch Umbilical Tower, built into the Mobile Launch Platform as a permanent accessory. As the stages took their places, the tower's nine arms reached out to them, and technicians hooked up connections ranging from electrical monitor lines to large-diameter fuel hoses. The stages would remain empty here, as would the fuel lines running up and down the 398-foot-tall Launch

Saturn 1B and Saturn V first stages
being completed at the Michoud
Assembly Facility in Louisiana. 1968.

Umbilical Tower. The platform would be plugged into the necessary reservoirs at the pad, where the fuel and propellants would be delivered.

A great window occupied the center of each end wall of the VAB, each forming a huge translucent rectangle made of hundreds of individual panes. Shafts of light fell through these windows into the cathedral-like space inside. The sounds of voices and crane operations were swallowed in this vast vault. The roof soared dizzyingly above, and the white and black hull of the Saturn V showed through the latticework that enveloped it.

Every stage of the Saturn V was impressive in size and achievement, but here in the VAB all the elements came together for the first time. As the crews worked intently on their myriad tasks over the days and weeks, the separate components from disparate origins were fused and metamorphosed into one gigantic instrument that was Apollo 11. Mechanically joined, electronically integrated and finally topped off with the delicate spacecraft, the total assembly struck awe into the viewers it dwarfed so completely.

The extraordinary nature of the Apollo project cast an aura over all its parts, an influence remarked on by many workers involved throughout thousands of manufacturers and laboratories across the nation. Preparing a part of the system that would carry human beings to the moon was profound. It inspired an effort toward perfection. To have touched the part of a moon rocket that would make the journey left, for many, the lingering sensation of having touched history.

The entire creation and assembly process culminated in the heart of the VAB. VAB technicians, from crane operators to engineers, felt the importance of their work deeply and could not have been more committed or dedicated if they had been installing the stained glass window in Notre Dame Cathedral, a task that might have been undertaken by their forebears. The magnitude of effort was comparable. Both monuments expressed the power of shared faith in something beyond the ordinary, in a purpose beyond simple explanation. In both cases, the height of the achievement would reach beyond simple measure.

ROLLOUT

The early Saturn Vs were awesome, but the one for Apollo 11 was even more so because it would carry the first lunar landing mission, and so accomplish Kennedy's challenge. When the last bolts had been tightened and Apollo 11 was ready, the great door of the VAB opened. Because the doors were measured in acres, the process took hours. Seven vertical leaves rose upward one by one. Horizontal doors below parted and drew back on tracks, until there was a great inverted "T" of open space gaping 456 feet tall and 152 feet wide. Through this portal the Saturn V would emerge.

Underneath the gray platform that held the launch tower and the moon rocket was the massive, tank-like crawler, which rose until it carried the whole weight of the assembly on its back. The driver aboard the crawler sat at the controls of a machine that could carry all of this at once: the giant platform, the tower, the stacked moon rocket.

They called the event a "rollout," and it was such that VIPs attended. The appearance of the great black-and-white moon rocket from inside this cathedral workshop was awe-inspiring. The stately pace of one mile per hour as the crawler steadily ground its raw horsepower against the load seemed fitting for its monumental grandeur.

People stood on the roof of the VAB, looking down as the colossus

→ **Apollo 11 lifts off toward man's first lunar landing. July 16, 1969.**

emerged. Technicans rode on the deck. On one of the first trips, Guenter Wendt took his station at the peak of the launch tower, communicating with the crawler driver about vibrations and making sure that all went smoothly. The emergence of a 36-story skyscraper from a building large enough to house four such spires was a statement of magnitude such as the world has not seen since.

The Apollo 11 spacecraft was destined for a round-trip journey of almost half a million miles. But first the Saturn V had to make it to the pad. The distance was just over three miles but the journey would require titanic effort, and the monument would carve its passing into the very face of the earth.

TRANSPORT

In the words of its present master, the crawler-transporter is "six million pounds of walkin', stompin' hell!" The crawler chief at KSC is Thurston Vickery, a strapping engineer who seems to stand little shy of seven feet tall, and seems taller yet when he speaks. Not just anyone can handle work on the crawler team. Suppose a tread breaks a link, and needs a replacement? Vickery has to be able to count on individuals who are not going to faint when they find out that the links weigh a ton apiece, and that the work has to be done immediately, wherever the crawler happens to be when the break occurs—it can't be towed into the repair shop. Vickery weeds through potential applicants by testing their reaction to a departmental "gotcha." "We had a six-foot crescent wrench made that I show to the new hires," he says, "Just to scare 'em. I bring 'em out here under the crawler, and I let them see that wrench. When they notice, I look 'em in the eye and tell 'em, If that looks big to you, *then you're in the wrong department.*"

The crawler's oversized parts look like they would just about need a wrench that size. The mufflers underneath the crawler are the size of Lincoln Continentals.

"We drive over rocks," says Vickery, "and when we get through, they're *sand.*"

The unwarned visitor might dismiss this as Texas-style hyperbole, but if you get special clearance to go out and look at the widely spaced twin-lane crawlerway, you'll see it surfaced with Tennessee river stone. The pebbles are nice and round like stones you'd find in a decorative fountain. They work like ball bearings to let the crawler move through the curves without peeling the pavement right off the ground. As the crawler moves past you like a mechanical glacier toward the pad, it recedes with its huge tread links grinding upward in the rear. If you kneel down to see the trackway it is making in the stones, you will see that the rubble surface is flat where the treads were only moments ago. There is a perfect impression of

The immense weight of the Crawler-Transporter leaves tracks as it grinds rocks into sand to mark its passing. 2004.

each tread link, left like a fingerprint. The stones are smashed.

A crew of spotters always walked along with the crawler, keeping an eye and an ear out for any problems, such as cracked tread shoes or noises that didn't sound right. The road they traveled is 131 feet wide, about the size of an eight-lane highway with a median gap. This crawler route was the highway to the moon. Its engineers built it to hold up under a load of 18 million pounds: the 12 million of the Saturn V and its launch tower/platform, plus the 6 million pounds of the crawler itself—that's 9000 tons. The foundations of the road ran, on average, seven feet deep. The trip down the crawlerway from the VAB to Pad A was 3.5 miles and took about six hours. The maximum loaded speed of the crawler was one mile per hour, but even today that speed is rarely hit while under load.

Vickery looks askance at the mere suggestion. "Flat out? This is no place for drag racing! Besides, the engines get to caterwauling something terrible at that speed." The driver's cabin has a remarkable speedometer on the dashboard that goes up to 2.

Imperfections in the roadway did not tilt the Saturn V on the crawler. The crawler's sensitive leveling system kept the needle tip of the Saturn V from moving more than 6 inches in any direction, and this level stability was maintained even when the crawler hit the ramp that led up to ground zero at the pad.

THE APOLLO PADS

The engineers prepared the area for the Apollo pads by heaping half a million cubic yards of sand and shell over the pad site, using material the canal makers had dredged up from the Banana River. The 1.5-billion-pound burden settled and crushed the ground below it into a more stable foundation. When it had sunk about four feet, the mound was cleared away and the construction was begun on a firm base.

At heart, the moon rocket pad did the same thing that a Redstone pad did. It sat the rocket up above a flame deflector and offered service connections. A mobile gantry tower gave technicians access and then moved out of the way before launch.

The blast plate underneath the Redstone had been a little metal pyramid. The flame deflector for the Saturn V weighed 650 tons and stood over four stories high. This height explained the long ramp that led up to the pad; the rocket had to be brought up to stand over the deflector. Its concrete surface was made with volcanic ash, but even this would not fully withstand a Saturn V blast. In view of the potential destruction, the deflector was on rails and could be pulled out and exchanged for a spare waiting on standby.

The deflector lay in the center of a flame trench that contained and carried the split blast. The fire shot in two opposite directions—out through the center of the approach ramp, and out the back of the pad. The trench was a 58-foot-wide, 42-foot-deep pit designed to withstand forces beyond most people's imagination. The vast size of the space now conjures a sense of the volume of the coming blast, which filled this channel and flowed like a river through it. The walls in the flame trench were lined with the kinds of bricks they use to build kilns, and they now bear heavy scars of previous incinerations. It is a burnt, blackened hell of a place down there. As a visitor, your body picks up on all the clues to the monumental danger that this pit harbors, and in spite of the privilege of visiting, you don't want to be there very long.

At ignition, the sonic energy alone could easily kill you, even if you

ran to the side before the flames washed through. The blast shot out of those engines at over four times the speed of sound, causing a rapid succession of mini sonic booms that sounded from miles away like staccato firecrackers.

Out of respect for the power sitting above those five engines, every Apollo facility that could be built away from the pad was built *far away* from the pad, so that a catastrophic explosion would not wipe out our chance of making it to the moon on time. The major structures sat at the edge of the blast zone.

The pad was surrounded by a large, fenced-off octagonal yard. The big tanks for kerosene were on one side, those for liquid oxygen on the other – as far apart from one another as possible. Stored underneath the pad were huge tanks of compressed nitrogen, and all over the site were the various chemicals, liquids and gases.

The setting was complete. Despite the size of the vast facilities and their tremendous magnitude, the coming event would dwarf it all.

9

MOONSHOT

T HE LAUNCH CONTROL Center for the Saturn V moon rocket looks unlike any other blockhouse ever built before, at the Cape or elsewhere. The building's relatively safe distance from the pad meant that it did not need to be a fortress, and its architects and engineers were free to design with inspiration in mind, instead of designing another squat fortress, pillbox, or armored igloo; they took a radical departure from all precedents. More than any other facility at the Cape, this building represented the nerve center of the moon landing program, and its designers wanted it to reflect an appropriate "futuristic" spirit: clean, attractive lines, gleaming white surfaces, and crisp bold details. The building still looks distinctive and futuristic today, worthy of its purpose.

← Astronaut Aldrin leaves the lunar module
  to set foot on the moon. 1969.

Every Saturn V stacked in the VAB was assigned its own dedicated firing room in the Launch Control Center. Each firing room contained a complete set of equipment and high-speed computer links that would follow an individual moon rocket from its earliest inception, monitoring its assembly and checkout, and commanding its launch, getting interrupted only by the period of transportation on the crawlerway. As with the bays in the VAB, three firing rooms were fully equipped, and one was kept in reserve for future needs. Designed to support the predicted high launch rate in the future, the setup was a key part of the mobile launch concept that allowed NASA to assemble one or two moon rockets while another was being launched out at the pad.

An Apollo Saturn V firing room represented the complexity of launch operations at a scale never before seen. Some 450 consoles kept watch over the systems of the Saturn rocket and the Apollo spacecraft atop it, each processing prodigious quantities of data. In keeping with the spirit of the building's design, the entire area was laid out somewhat theatrically. The banks of primary technical monitors sat on the main floor. Stepped levels built from the main floor up to the highest areas following rank, giving launch leaders a view over their troops below. Glassed-in alcoves on the highest level provided quiet cocoons for the highest VIPs such as Debus and von Braun. Large main screens hung above the console floor, providing a powerful visual statement about their unity of purpose despite the myriad individual tasks that kept each technician closely tied to their consoles. A countdown milestone indicator illuminated each critical stage advancing to liftoff. Although not absolutely essential, such touches affirmed the drama of the situation and were the kinds of gestures that fostered dedication and team spirit. These large visual elements also helped communicate the purpose of the room to television viewers to whom a mass of identical consoles would not have spoken so vividly. When this Launch Control Center setting appeared on the news, the American public saw exactly what this crew was working toward, at a scale that matched the grand scope of the imagination involved.

Everyone faced away from the long window that looked out toward the launch pads. Their backs were to the distractions of the view outside, and their heads toward their monitors. Above them were the main screens if they needed a glance to remind them of context or purpose. Once the team had done their jobs and the rocket had cleared the tower, Mission Control in Houston would take over from Launch Control in Florida. Then the control center technicians could turn around en masse and watch the launch thunder up from the pad. The sight would be an incredible reward.

APOLLO LAUNCH

SOME OF THE journalists who had been covering the manic progress of the Space Race from the beginning confessed to being jaded by the time Apollo 11's prelaunch hours arrived. Throughout the country naysayers, jokes and searching editorials decried the expense, denouncing the pointlessness of sending men to the moon. Some of it was the same blind criticism that had been dished out at the "damn fool" Wright brothers for their useless airplane experiments. In other respects, however, some of the critics seemed to have a point. The Reverend Silas Abernathy, for example, a crusader for the poverty-stricken who led a march to the gates of the Cape, saw only the heart-rending injustice of such public expense when there were Americans in need. Apollo was a uniquely visible federal program, and thus an inevitable target.

Opponents tended to ask "Why are we spending $24 billion on going to the moon when we could be spending it on program such-and-such?" While this sounded compelling, this was a false dichotomy. Given the complexities of federal budgeting in Congress, the cancellation of Apollo would not simply provide funds to the variety of programs matched against it. Every program had to compete on its own. The real question Apollo raised was "If Apollo is about to prove that today Americans can do literally *anything* we put our minds and efforts to…why have we not chosen to put the same dedication into solving our social and other problems?" That was not a question NASA's engineers could give an answer to, but one for the nation's conscience to contemplate.

Apollo was set in motion to prove a political point, and the moon continued to hang in the sky more as a symbol than an astronomical body, no matter how many equations about its gravitational properties were factored into NASA trajectory plots. Poets publicly lamented that astronauts and moon probes would transform the traditional beacon of lovers' inspiration into something dull and prosaic, a world of dust and stone. On the brink of a tremendous change, they could not see that the redefinition would be one of such profound human meaning that even the poets would be beggared for words to express it.

Wernher von Braun spoke the night before the launch. He observed that what was about to happen would change the world just as had Columbus' footfall upon a beach unknown on European maps. "When Neil Armstrong steps down upon the moon…the ultimate destiny of man will no longer be confined to these familiar continents that we have known so long." There would not be an edge to the map any longer.

Thousands of people camped on the beaches and roads near KSC to watch the Apollo 11 liftoff.

Half a million people had made their way to the Cape, to be present at the moment, to witness history. The grounds around the Space Center were filled with the pilgrims' tents, campers, trailers and sleeping bags. People crowded out to the very edge of the water where jetties stretched closer to the launch site. Binoculars, tripods, telescopes and expectant faces of every description peppered a sea of humanity humming with anticipation.

On the night before the launch, Apollo 11 stood plainly visible to the crowds camping along the Banana River. This was no Soviet or Chinese secret hidden away in unknown wastelands; Apollo 11 stood open before the people of the world. Searchlights bathed the white spire of the Saturn V in an incandescent gleam, the beams streaming up like the reflected glory of heaven. The rocket stood ten miles away over the water, yet it was large enough to see clearly. With binoculars, you could see the letters "U," "S" and "A" running down the first stage, and then count the stripes of "Old Glory" on its flank. As an American citizen, you felt that this incredible craft was somehow related to you. That flag was the same flag that flew on your porch. Here it was on a triumph of engineering that looked a perfect symbol of the reach for the stars with its shimmering whiteness, its slender taper, its fine point, and its strange, stark black markings.

If you had clearance to enter Kennedy Space Center the next day, you made your way past miles of people packed more tightly into the place than ever before. The audience lined the shoulders of Highway 1 and crowded every bit of viewing ground all the way down to the river. The mood was reminiscent of the Fourth of July, and the coming fireworks of the liftoff would not disappoint.

Reporters at the press site numbered 3,493. Fifty-five other countries had sent media emissaries to watch the launch. The assembled VIP guests included former President Lyndon Johnson, who had as both vice-president and president supported the space program legacy of his slain predecessor John F. Kennedy. Von Braun's mentor Prof. Hermann Oberth sat in a place of honor, along with Rudolf Nebel, the man who had established the Model Rocket Field in Berlin, home to von Braun's earliest efforts.

The Saturn V was the fulfillment of von Braun's vision, but it was also a fusion of creative energies far beyond his influence. America had not only the resources and the leadership to give wings to von Braun's space exploration dreams, it also had the collaborators who could meet the challenge of von Braun's genius and create together the finest piece of engineering the world had ever seen—the Rocketdyne engines, the Grumman lunar lander, and all that went into the brilliant work. The Saturn V was von Braun's baby if it was anyone's. There were those who said that the coming footprints on the moon would most truly be his. But since Peenemünde, von Braun had understood that a great rocket could only be created by a great team. It was this realization that made him fight to keep his group together, through their migration from Germany and through the vagaries of their fortunes as Americans. The Apollo team was far larger than even the rocket team's extended family that had built the parts in hundreds of companies across the country. In all, some 400,000 people had participated in the program. Apollo had reached this point not only because of the rocket team and its commander, and not only because of the leadership of John F. Kennedy, who had done so much to forge the dedication to the lunar landing program, but because it was carried by the strength of the American economy and the abilities of its workforce. In total, Apollo represented the power of America to rise to a challenge worthy of its national effort.

Neil Armstrong and his crew had breakfasted in the Manned Spacecraft Operations Building. Afterward they suited up with the help of several technicians, quietly skilled twentieth-century pages armoring space knights. The three for the moon took the elevator down to the ground floor and walked down the hallway, waving to reporters before boarding the van that took them out to Pad 39A where their dragon awaited.

Earlier, the launch team had thrown the valves that sent the propellants out from their tanks through the conduits across the green octagonal yard into the pad at its center. The blood flowed upward through the veins of the great orange tower and out across eight of its nine arms. It poured into the Saturn V... and poured... and poured... until the entire column was no longer empty, but nearly bursting with oceans of what was like liquid dynamite.

The fully-loaded Saturn V that greeted Neil Armstrong, Buzz Aldrin and Michael Collins upon their arrival that morning had come alive. Its metal hull creaked and groaned. Its connections hissed. Vapors floated about. Ice caked the hull where the cold of cryogenic oxygen reached through all the layers and turned the moisture in the air to frost. The Saturn V strained at its moorings. The phenomenon that von Braun's rocketeers had known ever since their early efforts had now come to pass on a titanic scale: there was a ghost in the machine, a coiled dragon, the sum of countless thousands of individual works transmuting effort into this machine's life and power.

The astronauts rode the elevator up the tower as stretch after stretch of the Saturn V hull slid past them. It just didn't end. As he rode the elevator, Michael Collins thought about the rumors "the whisperers" had told him, of the second stage being *the weak link in the chain*. North American Aviation had built the command module that killed Gus Grissom and his crew. North American had also built the only Saturn V stage to catastrophically rupture during tests. They further built the Apollo 13 service module that, unknown to Collins, would soon explode in space. The only major components of Apollo ever to fail disastrously would all be built by one company. North American had put together the gigantic second stage that Collins watched sliding past from the elevator. One could wonder: was it supposed to make this much noise?

There was no room for terror in the midst of the proceedings. Everyone involved went through practiced motions like clockwork. The astronauts reached the level of the ninth arm of the launch tower and walked across the caged gangway to board their spacecraft. Inside the white room, the final threshold, Guenter Wendt offered Armstrong an oversized "key to the moon" as his parting gift. The astronauts had to shout to talk through the glass bubble helmets they wore. Wendt and his

team methodically loaded the three into the cramped space of the Apollo capsule. Armstrong's wife Jan and his children, watched from a boat not far away.

When the closeout crew had withdrawn, Armstrong used a special key to arm the escape tower above his capsule. This rocket was powerful enough to pull the entire capsule forward, away from an exploding Saturn V. The escape tower alone packed far more punch than Alan Shepard's entire Redstone rocket had done only eight years ago. Such was the dizzying rate of American progress in those days that what had powered an entire space vehicle now powered just a small safety accessory on the Saturn V.

An accidental triggering of the escape tower while either the astronauts were boarding or the closeout crew were still in the white room would have resulted in many fatalities, so the system was not live until Armstrong turned the key. For John Glenn, they had cleared the gantry just to turn on the system, so fearful were they that it might go off accidentally. Turning the key in Apollo 11 was just one line in Neil Armstrong's long checklist.

The great Saturn rocket shown under development at MSFC dwarfs the earlier Juno (rear left) and Mercury-Redstone rockets. 1961.

Seen from across the water, sparks appeared underneath Apollo 11, followed immediately by the billowing of white clouds from out of the flame trench. The flood of water from the deluge system was meeting the dragon's breath. For six seconds the thrust built and the clouds rose, then the clamps let go.

From the VIP stands, observers saw the entire vehicle slowly move upward, rising above the obscuring white clouds, and then sensed that the sound was coming. You could feel it traveling toward you, racing across the water and the marsh like a horseback army until it arrived like an earthquake. The Saturn V had cleared the launch tower by the time the sound hit. The ground shook, your clothing shook around you, the vibration shot clean through you. Fish jumped out of the water. The light of the rising Saturn V's flaming tail was surprisingly and painfully bright, like an arc-welder's torch. From over three miles away its force and its heat could be felt. Its 2400°F flames blasted outward at Mach 4, hammering at the earth, at the Cape, and at the air in crackling elemental fury. The engines thundered through an ungodly 15 tons of propellants per second, forcing the torrents into total screaming conflagration. It was as if the sun was erupting through a volcano.

Three men inside the Saturn V cockpit endured battering vibrations as they left their world behind. Below, the watchers marveled at the sight they had come to see as the rocket climbed into the heavens. "In a very real sense," Kennedy had said, "it won't be one man going to the moon. It will be the whole country, for it will take all of us to send him there." There was electricity in this crowd. Other crews had ridden the Saturn V, but this crew was different. They *were* going to the moon. *LIFE* reporter Loudon Wainwright, watching Apollo 11's spectacular rise with everyone else, called it "the ear-splitting, eye-smashing beginning of the greatest trip in history."

Apollo 11 did zero to sixty in four seconds, and it would only get faster from there. The five main engines howled with astonishing horse-power. The rocket began its "roll and pitch" program almost immediately after launch, the huge Saturn V steering with its four movable outboard engines to turn and lean into its rise, moving away from the tower for safety. As it rose it arced over, then headed out over the Atlantic. Two minutes and 42 seconds into the flight, the rocket was almost lying hori-zontal, and the kerosene-burning first stage was spent. Explosive bolts cut it loose, and the second stage flared into life and took over. This was exactly the maneuver that Bumper 8 had been launched to test, 19 years earlier, just a short distance down the coast. Here was the future, and this was the road to the moon.

The first stage, abandoned, would fall 42 miles and splash into the Atlantic. Von Braun had wanted to build recovery systems, since the main engines were built so well that they could have been re-used several times, but Apollo was about racing to beat a deadline, and adding re-usability was passed over in favor of getting the Saturns out the factory door.

The Saturn V drilled onward with its second stage, gaining speed, passing Mach 10, then Mach 11. Achieving orbit is not about reaching a certain height. You can rocket straight up to 117 miles, but you'll only fall right down again, like von Braun's V-2s over White Sands. Achieving orbit is about reaching a certain velocity, and that's why the Saturn V leaned flat over so soon after launch, to accelerate toward its orbital speed of 17,000 statute miles per hour. At that velocity you are traveling so fast that as you fall you circle around the earth. You are still "falling" the entire time, and

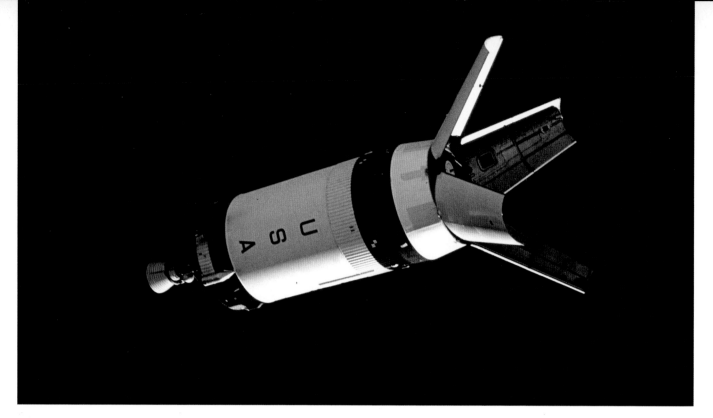

↑ The Saturn V third stage, the Lunar Module Adapter, was photographed with its arms in the fully open position during Apollo 7. 1968.

that is what orbital "weightlessness" feels like. Astronauts maintain perfect concentration and pilot their precision craft while feeling, the entire time, as if they are in a skyscraper elevator whose cable has just been cut.

The second stage handed over to the third stage, which needed only a fraction of its fuel to achieve the low parking orbit. It was then shut down while the spacecraft's systems were checked out. Only when Mission Control gave the go-ahead did they restart the third stage. It blasted Apollo 11 out of Earth orbit and sent it on to the moon.

En route to the moon, the pilot Collins did a balletic maneuver to pull the lunar lander out of its garage atop the spent stage. The conical garage opened like four flower petals, exposing the delicate lunar module crouching inside it with folded legs. The command module nosed into the lander's topside docking hatch, and then with the triggering of a release, springs gently pushed the docked spacecraft away.

The three stages of the rocket had done their job, and it was now up to these two spacecraft to travel all the way to the moon, accomplish the landing, and then bring the astronauts back.

Neil Armstrong's first step on the moon on July 20 was watched by some 400 million people around the world, and perhaps listened to by another 600 million on radio, as he said "That's one small step for [a] man, one giant leap for mankind." Buzz Aldrin soon joined him, and the two saluted the American flag for one of history's greatest images. Armstrong spent just two hours and 41 minutes on the gray, dusty surface of the moon, because NASA wanted to proceed with extreme caution.

→ In this iconic image, astronaut Harrison Schmitt plants the American flag on the moon. The earth appears in the sky just above the flag. 1972.

Every mission that returned to the moon would stay longer and go farther. This was the mandate of exploration: go higher, go farther, and always add new territories, new horizons, to the human experience.

NASA sent seven missions to land on the moon. Six made it. Due to a communication error by North American Aviation regarding an oxygen tank, Apollo 13 suffered an explosion during the long coast to the moon, but the crew was successfully brought back using desperate measures, determined effort and improvised equipment that stretched the remaining oxygen.

Apollo 12 demonstrated pin-point landing capability, setting down just 600 feet from one of the Surveyor soft-lander probes. Apollo 14 tried, but failed, to climb to the edge of a large crater rim. The final three Apollo missions were the most elaborate and adventurous, using an electric car called the lunar roving vehicle to carry the astronauts miles across the surface. With increasing experience, NASA grew more daring, and the hotshot pilots attained more difficult and interesting landing sites. Apollo 15 landed in the middle of a valley on the far side of a range of mountains, with a canyon just beyond; the astronauts worked their way up the flank of the mountains on one side, reaching breathtaking views. Apollo 16 explored fairly nondescript terrain in the highlands, but Apollo 17 set down in the middle of a spectacular plain ringed tightly by steep hills. Over three days, astronauts Gene Cernan and Harrison Schmitt were able to range from one side all the way to the other, collecting a complete range of geological specimens. Between the several missions, the Apollo moonwalkers returned evidence that enabled scientists to unlock the secrets of the moon, in turn shedding astonishing new light on the formation of the earth.

Two more missions beyond Apollo 17 were on the books. They might have ranged to even more fantastic settings, possibly on the far side of the moon, or explored the mysterious and inexplicable glows that sometimes appeared in certain lunar locations. However, the final Apollo missions were cancelled to please the Nixon White House. President Richard Nixon had declined to attend the launch of Apollo 11 and would oversee the burial of the program so thoroughly identified with predecessor John F. Kennedy. The Saturn V rockets were already built, the main expenses already incurred, but the rockets were scrapped anyway. Only one Saturn V would be spared—this instrument would make possible a remarkable coda to the Moon Race.

→ A hero's welcome greets the Apollo 11 astronauts as they parade through New York.

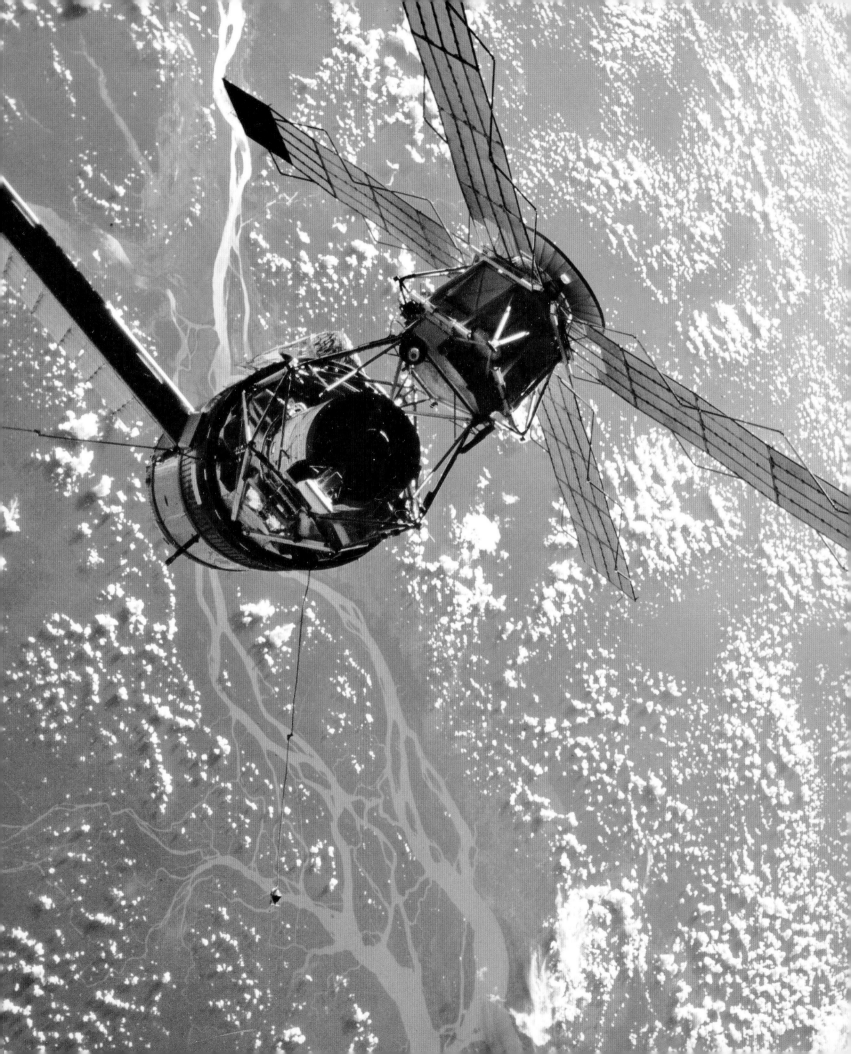

SKYLAB
AND SOYUZ

10

THE FINAL ACCOMPLISHMENT of the Saturn V moon rocket was to launch an entire space station into orbit in a single bound. Wernher von Braun had the idea for building a space station out of a spent rocket stage. He considered the "type IV" stage and thought, once it is empty of fuel, you have a huge tank floating around up in space...why not fill it full of air and fit out the interior as a space station? The hard part was getting the station's hull up there. Once you had the container, you were halfway home. He sketched the idea on a scrap piece of paper in the mid-1960s.

← **The Skylab space station passing over the Amazon River Valley. 1973.**

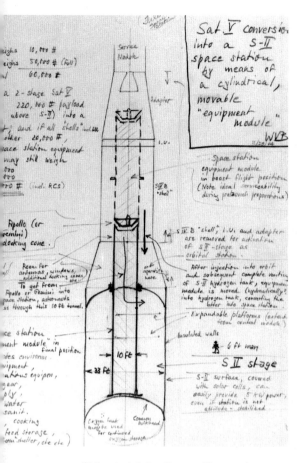

This sketch by Dr. von Braun shows his early concept for developing a Saturn V second stage into a space station. 1964.

→ The Skylab 1 orbital workshop is being mated to a Saturn V in the VAB prior to launch. 1972.

I N 1973 THE 100-ton Skylab space station was being stacked on top of a Saturn V rocket in the VAB. The idea had evolved from fitting out a spent tank in orbit to fitting out one "dry" on the ground. Building the station on the ground rather than carrying out multiple component launches saved fortunes and allowed the station to be launched in one efficient piece. It would hold an incredible amount of usable volume, as well as instruments for observing both Earth and space, and multiple decks for carrying out medical and science experiments. All this was possible because of the power of the Saturn V. No other rocket could shoulder such a load, but when the cancellation of the Apollo 20 moon landing mission made one available, von Braun's conversion station idea went from a Saturn IB launch to a Saturn V, and suddenly tremendous economies were available.

Skylab's core was, just as von Braun had imagined, the third stage of a Saturn V. Ordinarily, this was the stage that blasted the Apollo spacecraft out of Earth orbit and sent it to the moon. If you weren't going to the moon, all you needed was the first two stages, and the third could become payload. This one had become Skylab, with several modules added to it to increase its capabilities beyond merely housing astronauts. Skylab would be a space station with a purpose, and it would send back reams of new data about the earth below and the sun above, in addition to results from zero gravity experiments of the kind that have since been endlessly repeated to explore medical, bioscience, and material properties. Skylab pioneered.

The buffeting of the supersonic slipstream during launch tore off Skylab's outer shield and one of its folded solar power wings. Its first two crews had to make repairs, carrying parts and special tools with them into space aboard their Apollo capsules. The repairs succeeded brilliantly, and Skylab became an amazing story of hardworking astronauts carrying out a nonstop regimen of experiments. They captured unprecedented views of the sun using the most sophisticated battery of telescopes ever orbited for the purpose, including a stunning eruption that stretched millions of miles into space. The second of the three crews who lived aboard Skylab set records for productivity that remain unmatched today.

The crews would ride to Skylab aboard leftover Saturn IB rockets. Built as backups and for test purposes, the rockets had never been used. NASA dusted them off and topped them with brand new Apollo capsules to serve as the crew rockets for Skylab. Pads 34 and 37 had been built to launch the Saturn IB, but keeping launch pads active cost a fortune, so NASA worked out a more economical way to leave those pads inactive and convert the active Apollo backup pad, 39B, to serve the Saturn IB. The IB had been a big rocket in its day, but to sit at the Saturn V's table it

Prior to the launch of Skylab, the Saturn 1B is perched on the "milk stool" launch platform to allow the 1B to use Launch Complex 39B which was developed for the much larger Saturn moon rockets. 1973.

needed a high chair. NASA built a "milkstool" to raise the smaller rocket up to a level where it could conveniently drink from some of the hookups already built into the Saturn V launch tower.

Launch Pad 39B had been a quiet and little-used backup pad for most of its life. During the lunar missions, it dispatched only Apollo 10. But it served well to launch the first Skylab crew shortly after the launch of Skylab itself. Having more than one active pad allowed NASA to stage close-succession flights like this and also made a rapid rescue launch possible.

During the Skylab program, the Cape had on standby a rescue kit that could be installed into an Apollo capsule in case the astronauts in orbit needed help. The rescue capsule could fly into orbit with two crew-

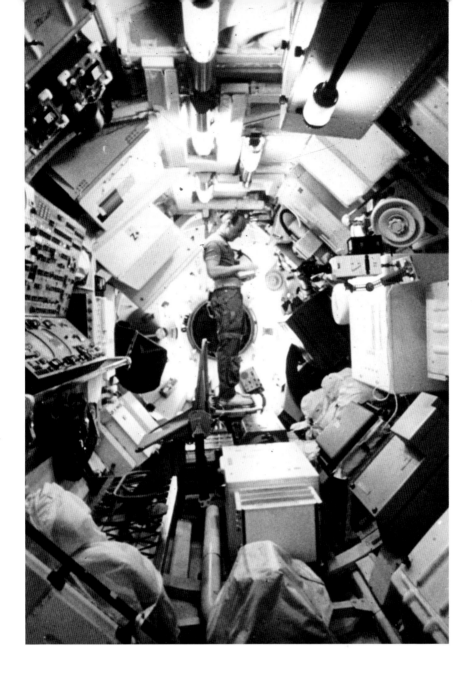

Astronaut Charles Conrad, commander of the first manned Skylab mission, goes over a checklist during training in the Multiple Docking Adapter. 1973.

men and carry back all five people. NASA does not maintain such active standby rescue capability today. The space shuttles do not launch with a second shuttle available for rescue on short notice, a factor that contributed to NASA manager Linda Ham saying, "there's nothing we could do about it" while *Columbia* orbited with suspected damage to its heat shield system. Even the old Skylab rescue rocket would have been able to fly up, survey the situation and confirm the damage, allowing NASA to work a solution. Lack of standby rescue launch capability also means that the International Space Station must operate on a skeleton crew too small to accomplish effective work. But during Apollo, standards of crew safety were higher. The rescue rocket was almost scrambled when one of the Skylab mission command modules appeared to be developing trouble with its maneuvering thrusters, but the problem was solved. The important thing was that the rescue capability was there.

Skylab was an exciting follow-on program to Apollo and seemed to be the beginning of a new era in space. The station delivered a bonanza of new images of the earth and the sun. Many of the kinds of experiments that have since been repeated at length aboard the space shuttle and the International Space Station were conducted for the first time aboard Skylab.

Skylab also gave NASA some of its most daring "right stuff" astronaut adventures of all time. The first crew flew carefully around the station to survey the damage to the solar wings. They found that one was missing, the other jammed partly open. In an effort to fix it, one astronaut held a pole with a hook and hung out of his capsule while his comrade held his feet, while the pilot brought him in close enough to yank at the wing with brute force. The station felt his pulling and tried to counterbalance the effect with its powerful gyros. The whole station fought and tugged away from the astronaut's grasp. The wing was jammed too tightly by a metal strap to open up this easily, but they would fix it by cutting the strap on the spacewalk. Still, this was the kind of "seat of the pants" flying and daring that space adventure was cracked up to be.

The only problem was that the public never appreciated Skylab because there were no pictures. The program was rather poorly photographed. Despite all the good work involved, there was no Neil Armstrong along to snap iconic images of it all. The lack of stellar images has contributed to the underappreciation of Skylab in history. It deserves better. It was the first and only American space station, and the best one ever flown. The first version of what became the International Space Station (ISS) was ordered by President Reagan in 1984 and was to be completed within 10 years, but by 1993, when President Clinton ordered that it be constructed in cooperation with the Russians, not a single piece of hardware had been placed into orbit. After more delays, the first elements, one Russian and one American, were joined in space in late 1998. It was first inhabited in 2000, so far over budget that NASA had lost count of the billions, as incoming administrator Sean O'Keefe discovered to his astonishment and chagrin soon after he took office. Compared to the ISS, Skylab was definitely the "right stuff." It was the last hurrah of NASA's golden age.

RUSSIAN RENDEZVOUS

AFTER YEARS OF tight security at the Cape, Russians penetrated the Space Center installations right into the heart of the VAB, and were able to examine everything. This was largely due to the set of VIP badges and security passes the Russians possessed. These visitors who might have triggered a military response a decade earlier had, in 1975, a cordial NASA escort. Times had changed, and NASA was preparing for the very first international space mission. American and Soviet spacecraft would launch in their home countries to rendezvous and dock in space. After international competition had been the very engine of the moon race, would space become the arena of cooperation? It was a bold idea. No one

This artist's conception depicts the first international docking of the U.S. Apollo spacecraft and the U.S.S.R Soyuz spacecraft in orbit. The actual docking took place successfully on July 15, 1975.

knew exactly where it might lead, but it would certainly be a new adventure for Apollo's last bow.

Skylab's rescue rocket still waited after the last mission came home. This fine piece of hardware escaped being scrapped to carry a special crew on Apollo's final mission: a space rendezvous with a Russian Soyuz craft. A Russian and an American "standard" space capsule, first piloted into space in 1967 and 1968 respectively, would be sent up. Our larger capsule would carry three men, theirs just two. Talks between U.S. and Soviet space officials on the possibilities of cooperation and the exchange of data and lunar samples had been going on ever since 1962, but the agreement for the rendezvous had only been settled with Richard Nixon's signature in Moscow on May 24, 1972, before the last Apollo moon landing had even taken place.

They would call this the Apollo-Soyuz Test Project (ASTP) or, if you were in the Soviet Union, the Soyuz-Apollo Test Project. With superpower prestige on the line, great pains were taken to treat both parties as equals in every respect. Aboard the docked spacecraft, the astronauts would speak to the cosmonauts in Russian, and the cosmonauts would reply in English.

Seven hours after the Soyuz fired into orbit from its launch base on the wasteland steppes of Kazakhstan, NASA's countdown hit ignition at the Cape. In Firing Room 3 of the Apollo launch control center, 440 people sent Apollo's final mission up from the milkstool on Pad 39B; it was July 15, 1975. After two days of orbital maneuvering and preparation, the two ships closed in to dock.

THE INTERNATIONAL DOCKING RING

THE SATURN IB customarily carried a conical adapter stage of the same type used as a lunar module bay on the Saturn V. On Wally Schirra's Apollo 7, the first manned Saturn IB mission, the "lunar module bay" was empty—in part because the Saturn IB was not powerful enough to lift the weight of a lunar module together with an Apollo capsule. On Frank Borman's Apollo 8 mission, the empty bay carried a ballast weight because the lunar module wasn't yet ready. During the Skylab missions, the bay had carried nothing. For the Soyuz mission, the adapter stage housed a unique module: a docking adapter with an American port on one side, and a new "universal" port that matched the one that the Soviet Soyuz spacecraft would bear on the other. Between the two ports of the docking module was an air lock tunnel, with just enough space for the men to float in together and shake hands.

The Apollo-Soyuz project was designed to give engineers on both sides the experience of collaborating to build hardware that was mutually compatible and to test it in flight. Cooperation in space also might lead to the possibility of future joint missions. The hardware at the heart of the mission was the international docking ring. This was the mission's primary engineering goal: to develop a standardized, universal docking system. A standardized design would greatly increase potential future options for cooperation. International rescues could even be mounted, since there would no longer be need for emergency design on the key points of compatibility between spacecraft. The project succeeded with the device they built for Apollo-Soyuz. Today the docking ring's basic three-flapped design can still be seen in all the ports of the International Space Station.

Shortly after reaching orbit, the Apollo command module pilot had spun his ship around, nosed in to dock with the module in the garage, and pulled it away from the booster stage. Now they were coming in to the rendezvous.

Like the Apollo program itself, the intention of this project was purely political, but science got to come along for the ride. The political side of the rendezvous dominated the news coverage, but, in fact, the docking

American astronaut Thomas Stafford and Soviet cosmonaut Aleksei Leonov meet in the hatchway leading from Apollo Docking Module to Soyuz Orbital Module. 1975.

module served as much more than a plumbing joint. The space inside was fitted out as a small space station, with a variety of instruments and experiments for the astronauts to run. For two days the crews carried out experiments, toured each other around their respective spacecraft, shared meals and exchanged gifts. Commanding the Soviet spacecraft was none other than Russian space hero Alexei Leonov, the first man to walk in space back in March 1965.

Controlling the Apollo-Soyuz mission jointly, with both mission control centers in constant coordination, was a major challenge. Communications were more complex than in the past, too: just over half of the communications between the docked craft and the ground would not be picked up directly, as was normal, but would instead be relayed for the first time through a communications satellite in a fixed position over Kenya, in geostationary orbit.

After the ceremonial visit with the Russians, the spacecraft undocked. The highly maneuverable Apollo spacecraft executed fly-around maneuvers as close as 492 feet to the Soyuz while the green, ant-like Russian craft sat passively. The Apollo ship then precisely oriented itself with the

American astronauts Thomas Stafford and Donald Slayton in the Soyuz Orbital Module hold containers of Soviet space food. 1975.

Soyuz between it and the sun, creating an artificial solar eclipse in space. From the ground, you gain nothing by blocking out the sun's disk since its light still illuminates the atmosphere, obscuring the corona. In space, if you cover up the sun's disk, you can see the corona. The rare event of an eclipse, with all that it reveals about the sun, could be created on command in space, and this would be the first time to try it in this fashion. The experiment was an intriguing follow-on to Skylab's solar observation work. After a second docking to further test the procedure, the ships separated permanently.

The Russians would be home within 48 hours of concluding the joint mission, since the Soyuz spacecraft had a maximum mission time of only six days. With their larger ship and greater capabilities, the Apollo crew performed six additional days of experiments, and executed some remarkable precision maneuvering in the course of several of these.

The great space station Skylab was still in orbit at this time, but the ASTP spacecraft did not fly up to visit. With the splashdown of the American capsule, the Apollo era came to a close.

THE SHUTTLE

It was a strange combination—an audacious, large orbiter riding a hybrid rocket sled. One of the astronauts would later call it "a butterfly strapped to a bullet."

11

VISION QUEST

THE SATURN V was by far the most powerful rocket America ever built. Most rockets flying in the year 2000 — the Delta, the Atlas, and the Titan — had origins in the late 1950s and early 1960s, but had been steadily developed to greater power and efficiency over the years. In contrast, the fantastic power of the Saturn V was abandoned.

Richard Nixon ordered NASA to turn itself instead to the space shuttle, a new rocket that would have to be developed from the ground up, though its main engines would derive from the mid-stage engines of the Saturn V. Nixon's legacy for NASA was the end of astronaut space exploration, and a detour into a program that would cost far more than the reach for the moon. The resulting spacecraft could only circle the earth, incapable of reaching even as high as the two-man Gemini capsules had done in 1966. The compromises in shuttle engineering took the lives of 14 astronauts in horrifying accidents that also destroyed two $2-billion space shuttle orbiters. Along the way, shackled for a generation to the demands of Nixon's paralyzing legacy, NASA would almost forget the meaning of exploration. Yet it all began with a hopeful vision, as engineers sought a design for a large spacecraft that would be reusable and, over the long term, cheaper to operate than single-use vehicles.

← The prototype Space Shuttle *Enterprise* was used to test flight and landing behavior. 1977.

←← Space Shuttle *Endeavour* approaches the Rotating and Fixed Service Structures on Launch Pad 39A. 2002.

SHUTTLE ORIGINS

As with any rocket, the design of the space shuttle would determine its launch system needs. To Kennedy Space Center would fall the task of launching the shuttle, but the shuttle itself might take any of a number of possible forms. It might even take off like a plane, requiring a launch strip instead of a pad. KSC could not begin to prepare the proper facilities until the shuttle's design was finalized.

Today the space shuttle is a familiar shape, a combination of four elements: the spaceplane, properly called the orbiter; a large mustard-colored external tank carrying high-energy hydrogen liquid fuel; and two flanking white solid rocket boosters. The combination—and only the combination—is properly known as the space shuttle. The spaceplane by itself is, technically speaking, only the orbiter, and only the orbiters have names; for example, the surviving *Discovery*, *Atlantis* and *Endeavour*, and the lost *Challenger* and *Columbia*.

The separate elements of the shuttle stack are so different that they look almost improvisational and unrelated to each other, tied together by slender straps. The shuttle's genealogy explains this odd combination. The space shuttle, and its strengths and weaknesses, can be better appreciated in light of its conceptual origins, which stretch back to the 1930s.

VON BRAUN SHUTTLE

Into floodlights a delta-winged plane dropped, returning from space to glide to a landing at a U.S. runway. It carried a crew of eight and had just dropped off a cargo in orbit, while 50 million people watched on television. It was 1955, and the program was the television series *Disneyland*, on which Wernher von Braun was pitching his concept for a space shuttle. In an episode called "Man in Space," Disney and his animators dramatized the vision for America's exploration of space that von Braun had debuted to wide public interest in *Collier's* magazine a few years earlier.

Von Braun's shuttle was designed as a reusable system, both its booster stages parachuting back into the ocean for recovery, refurbishment and refueling. To avoid copyright infringement, the Disney version of the space shuttle featured delta wings instead of the canards and wide-swept wings of von Braun's shuttle in the *Collier's* paintings. The delta wings were suggested by von Braun's friend and fellow space popularizer Willy Ley. The Disney design was amazingly close to the shuttle the U.S. would eventually build.

X-15 REUSABILITY?

WHEN NASA WAS examining its options in the early 1970s, the closest thing to a space shuttle that had ever been built was the incredible X-15 rocket plane. Test pilots, including Neil Armstrong, had guided this wicked-looking black experimental craft to record-setting velocities up to Mach 6.7 (over 4,500 mph) and altitudes up to 67 miles — so high that the ship needed rocket maneuvering thrusters in its nose to steer in the space-like thin air. The pilots rated astronaut wings for these flights. NASA would adopt the X-15's "nose thrusters package" concept for maneuvering its shuttle in space.

The booster stage for the X-15 was an airplane, a B-52 bomber that carried it aloft under one wing, then dropped it at high altitudes. The main rocket in the X-15 then kicked in and shot the black ship up to the edge of space with acceleration so crushing that even the tough test pilots would admit they were glad when it was over. After the peak of its flight, the ship glided home for a landing at Edwards Air Force Base in California.

↑ In the 1950s Dr. von Braun worked as a technical advisor with Walt Disney (left) making television programs about space exploration. 1954.

↖ Dr. von Braun and Willy Ley (right), a popular author of books about space, hold a model rocket in the Disney studios. 1955.

Here was a real-world prototype of the space shuttle in many ways. The X-15s, "semi-space planes," were refurbished quickly in a NASA hangar and sent up for further flights, encouraging NASA that such quick turnaround would be possible with the space shuttle orbiter. Envisioning the shuttle as akin to the proven X-15 fueled NASA's dream of a rocket plane that would provide airliner-like access to space, without the months of ponderous preparation required for a one-shot Apollo-Saturn V mission. Ease of service and reusability had distinguished the X-15 program, and NASA forecast comparable results for its coming space shuttle.

DYNA-SOAR RIDE

IN THE 1960s, the air force wanted badly to get into the astronaut business, so they revived Sanger's antipodal bomber idea and developed the concept for a rocket-launched one-man spaceplane. The idea was to launch it over the North Pole, bomb the Soviets from high above the range of their air-defense missiles, fly around the world, and land just 100 minutes after liftoff. The rocket plane would return on a glide, using dynamic soaring, hence the project's quirky name Dyna-Soar. Reentry heat would be a major challenge for any plane-like shape, so the air force conducted a series of tests under the name ASSET in which Cape rockets sent up little nose cones in the shape of the Dyna-Soar's prow to test their responses to hypersonic reentry heating, temperatures so extreme that they would have melted even the titanium-hulled X-15.

Instead of launching the Dyna-Soar on a rail like a V-1 winged missile, the air force would follow von Braun's lead and position the little plane right on top of the most powerful rocket they had handy, a modified Titan II intercontinental ballistic missile. That alone wouldn't be quite enough power to lift the space bomber, so they would strap two large white solid rocket boosters onto the Titan's sides. Solid rockets were cheap and powerful, but comparatively dangerous and uncontrollable. They had been considered unacceptable for astronaut rockets, but soldiers could be expected to take such risks. The hybrid carrier rocket thus devised for the Dyna-Soar took the name Titan IIIC.

Dyna-Soar never flew, cancelled in favor of a military space station, which came very close to flying later in the decade. But due to a crucial NASA military compromise, the hybrid rocket configuration of the craft would live on in the design of the space shuttle. NASA wanted to send up its spaceplane on a piloted plane-like carrier that would return to base once it deployed the shuttle. Such a carrier might have been affordable for the small vehicle NASA had in mind, but the air force wanted a substantially

The X-15 rocket plane, shown here just after it was released from the B-52 that had carried it to the edge of space, provided NASA and the Air Force with much preliminary research for manned space flight. 1961.

larger shuttle that could carry heavy military payloads and fly the Dyna-Soar's strategic flight plan. Without military support, Congress would not approve the shuttle program. The larger "mil-spec" shuttle would cost so much on its own that its booster would have to be jerry-rigged from the cheapest possible materials, at the price of safety. Thus a basic configuration much like the Dyna-Soar's Titan IIIC booster would appear underneath the NASA shuttle, with a liquid-fueled center flanked by solid rocket outriggers.

NAVAHO CONFIGURATION

Von Braun's 1950s shuttle concept positioned its spaceplane atop its rocket booster, an arrowhead at the tip of the arrow. The Dyna-Soar had done likewise. But when it came time in 1972 to settle on the actual shuttle design, NASA chose instead to follow the configuration pioneered by the famously unsuccessful Navaho winged missile. The Navaho program had proven that it was technically feasible to mount a plane-like vehicle alongside a rocket booster.

The Navaho configuration chosen for the space shuttle places the shuttle orbiter in several kinds of danger. The delicate spacecraft is closer to the violence of the primary liftoff rocket blast than any other spacecraft in U.S. history. Any problems with the main engines, the solid rockets or the external tank occur right alongside the orbiter rather than safely behind it.

Given this especially dangerous position, the shuttle needed all the safety measures it could get. Every U.S. spacecraft had afforded its astronauts an escape system, the best being the Mercury and Apollo systems in which escape rockets could rip the nose capsule right off an exploding rocket and pull it to safety. Unfortunately, any comparable escape system was considered too expensive for the shuttle orbiter. Ejection seats on the earliest flights would allow only the pilots to escape, during only the first few seconds after launch; no provision for passengers on the lower decks was ever made. They would have to rely on the shuttle's perfect performance to survive. As one top Houston shuttle official claimed, the shuttle did not need escape systems since it was "built not to fail."

During the concept development, Apollo spacecraft designer Max Faget looked at the solid rocket boosters of the space shuttle and pointed out that any flawed joints in the hull might spit high-pressure flame during the solid rocket burn. The safer thing to do, he argued, would be to make the white solid rocket hulls in one piece, to absolutely prevent any burn-through. The one-piece hull would indeed be safer…but it wouldn't be political. A large one-piece casing could not be shipped long distances, cutting international contractors out of the bidding. In order to open the project to these distant contractors NASA was required to adopt a design with several separate segment joints.

The shuttle reached its final configuration under the supervision of NASA's Manned Spaceflight Center in Houston, which traditionally oversees astronaut spacecraft development. A reusable orbiter carrying three main engines would ride alongside a lightweight outboard "gas tank," flanked by two large solid-rocket boosters that would provide most of the liftoff power. It was a strange combination—an audacious, large orbiter riding a hybrid rocket sled. One of its astronauts would later call it "a butterfly strapped to a bullet." This design was the shuttle that we know today.

→ **This Navaho, launched into the air by three liquid-propelled rockets, was a forerunner of the Redstone and Jupiter rockets. 1957.**

TRANSFORMATIONS

Once NASA knew what the shuttle was going to look like, Kennedy Space Center could begin preparing facilities for the new vehicle. Generally, because rockets have such elaborate and specific support needs, a new rocket needs a new launch complex. Every time a new rocket came along, it required the building of a new pad, a new assembly building and a new blockhouse for launch control. This was one of the heavy expenses of progress in rocketry, and it was just part of the game. The space shuttle, however, would fly from reconfigured Apollo-Saturn V launch facilities, much like the Saturn IB had done later in its life for Skylab and Apollo-Soyuz. The savings in construction costs would be immense. The economical reuse of Apollo facilities was possible largely because Apollo had always built with an eye to the future.

Amidst the standing Saturn V facilities undergoing modification, the shuttle would get two major new facilities at Kennedy Space Center: a special landing runway and checkout hangars. These were new needs. Over the years, additional small facilities would be built to serve the space shuttle's needs more precisely, but nothing on the order of a new VAB or a new launch pad. Smaller new facilities would include buildings for solid rocket booster processing, shuttle main engine servicing and on-site shuttle heat shield component manufacture.

KSC began preparing itself for the shuttle while North American Aviation struggled to build the spacecraft. KSC cut the Apollo launch towers off the mobile launch platforms and used parts from two of them to build permanent umbilical towers at the pads. Workers cut new holes in the Apollo mobile launch platforms to accommodate the extra engines on the shuttle. Besides the hole in the center, there were new holes for each of the solid rocket boosters to blast through.

At the pads, a rotating tower structure was built, attached to the main tower. This was a new version of the leaning tower of Gemini, but it pivoted horizontally rather than vertically. It serves the purpose of the Apollo Mobile Service Structure, folding in to protect the shuttle while it is on the pad and providing a clean room for the loading of vertical cargo. It swings out of the way of the launch blast before launch.

The VAB got new work platforms custom-fitted to enclose the shuttle the way the original platforms had enclosed the Saturn V Moon rockets. The Launch Control Center was outfitted to monitor the specific systems of the shuttle.

SHUTTLE IMAGE

BEGINNING WITH NIXON'S original executive order of 1972 that brought the shuttle into life, there seemed to be a deliberate attempt to avoid the dramatic and inspiring gestures that had characterized Apollo. The shuttle program would have no poetic name—it would not be called the "Athena" spaceplane or any other such classical reference. In contrast to the heritage of Mercury, Gemini and Apollo, the shuttle would simply be called the Space Transportation System. The winged spacecraft of Von Braun's *Collier's* paintings often used to carry large numbers on their wings, asserting the proud and amazing fact that there were more than one of these massive spaceships, as well as allowing the public to easily distinguish between them. The space shuttles, however, would look anonymously identical. Only twice, by accident, were two of them ever photographed together. The entire group of orbiters was certainly never gathered together for a "hero shot" glorifying the incredible magnitude of the American Space Fleet for the public. Such gestures would not suit the character of the shuttle program. As a result of this anonymity, the general public would always remain vague on exactly how many orbiters NASA had.

In contrast to rockets with names like Atlas, Titan and Saturn, the shuttle's carrier rockets would have only bland, descriptive nomenclature—the external tank and the solid rocket booster. Inside the Launch Control Center, the grand main screens and the countdown milestone indicator were deactivated over time as "too dramatic." Technicians would be expected to keep their heads down. The glassed-in VIP balcony was painted over. In keeping with this approach, the new shuttle astronauts seemed to be selected for their anonymity. None would ever become household names unless catastrophes took their lives.

In these ways, the shuttle was positioned as a prosaically utilitarian piece of equipment rather than an instrument of heroics. Its aims were humble, the economical delivery of cargoes to low Earth orbit.

In spite of all the factors subduing the shuttle program, the fact remained that gigantic spacecraft were built and launched, carrying crews of up to eight into space. Instead of a burnt-up Apollo capsule drifting helplessly into the ocean at the end of its journey, the shuttle orbiter returned from space under the pilot's control, in grand majesty. From landing to launch, the shuttle was a major astronaut spacecraft system, and its operation contained unique dramas all its own.

THE SPACE SHUTTLE

12

THE ADVENTURE OF recovering, maintaining, preparing, assembling and launching a space shuttle orbiter belongs above all to Kennedy Space Center. The shuttle spacecraft has the ability to land in a few other locations if necessary, and the orbiters all periodically get shipped "home" to the Boeing plant in Palmdale, California, for major maintenance jobs. However the standard shuttle cycle occurs entirely at Kennedy Space Center.

Since the shuttle is re-usable, the cycle does not begin with launching. For KSC, a launch is the end of the line and the conclusion of their responsibility. It is at that point that Houston takes over with Mission Control, guiding the shuttle in space. For KSC, the space shuttle cycle begins with orbiter recovery, the decision to spin the shuttle around backward in space and fire its braking engines to drop it out of orbit. This retrofire happens on the other side of the planet from Florida, about an hour before a planned landing. Then the spaceplane spins back around nose forward, pointing slightly up, and begins to lose altitude. By this time the KSC team is already well into their preparations, and everyone involved feels the electricity of the alert: *Incoming.*

← *Discovery* being lowered into place to be mated to the External Tank/Solid Rocket Booster assembly. 2005.

SHUTTLE LANDING

GUNS GO OFF, their reports echoing across Merritt Island, startling birds away from the runway, just as they are designed to do. The salute signals that the orbiter is near, just as the spaceplane's distinctive double sonic booms announce its close presence above as it hurtles in across Florida. Windows rattle below its path. Only close to the end of its approach does this tremendous glider finally drop below the speed of sound.

The orbiter landing strip is one of the world's longest runways, stretching almost three miles through the lush Cape greenery. At 15,000 feet long and 300 feet wide, the runway is longer and broader than a standard jetliner runway, but comparable in size to research and development runways like those at Dryden Flight Research Center on Edwards Air Force Base in California, where the shuttle was landed in the early days of its development and test flights.

When coming in for a Florida landing, the shuttle orbiter homes in on its own special runway called the Shuttle Landing Facility (SLF). In the early 1970s it was imagined that the old Skid Strip out on the Cape would be improved to receive the shuttle orbiter, but NASA created an entirely new landing site especially for the space shuttle system instead, situating it closer to the VAB and the launch pads at Kennedy Space Center on Merritt Island. The facility opened in 1976, well in advance of the first operational orbiter launch in 1981.

The Shuttle's concrete runway features 50-foot asphalt shoulders on each side, and 1,000 feet of paved overruns at each end that can support the orbiter's 100-ton weight if necessary. In cross-section the runway is peaked like a roof, with a slope of 24 inches from centerline to edge for drainage. It is the job of a special crew to clear the runway of any foreign object debris in advance of a shuttle landing. This sometimes includes alligators that have crawled onto the runway to sun themselves.

NASA originally cut grooves into the runway surface to carry rainwater away and to improve traction for the orbiter on landing. The idea was to prevent any danger of hydroplaning. It was a sound safety measure, but the high traction caused excessive torque on the tires at landing and badly tore up the orbiter brakes. Further, the orbiter maintenance teams noticed that the grooves acted like a giant cheese grater, shaving tread off the orbiter's expensive tires so badly that one actually blew out at the end of the 51-D mission in April 1985. NASA ground the grooves off the touchdown area first, but in 1994 when it was clear that there was still tire damage, they just shaved the whole runway smooth.

Aerial view, looking north, of the 15,000-foot long runway at the Shuttle Landing Facility. 2004.

There is only this one shuttle landing strip at KSC, but it is called by different names depending on which direction you approach it from. It's called Runway 15 from the northwest, Runway 33 from the southeast. These numbers abbreviate the headings used by pilots to land in each direction—150 and 330 degrees. The approach direction will depend on wind conditions; you want to land a gliding orbiter into the wind, for the extra lift.

The orbiter often reenters the atmosphere somewhere over California, streaking across the United States as its heat-resistant tiles and other thermal protection layers defend its aluminum inner hull from the searing temperatures outside. About the time it crosses the Texas–Louisiana border, the spacecraft—now an aircraft—banks into two long S-curves to bleed off its high speed and help slow it down further. Around the time it crosses the Indian River the orbiter goes subsonic, and the commander takes manual control. The ship overflies KSC and the commander banks the orbiter into a great wheeling glide turn over the Atlantic, circling all the way around, back over the coast again, to cut across its own ground track as it comes in for the runway. It is as if it had taken a spiraling freeway on-ramp in reverse. They call the shuttle "a brick with wings," so rapid is its sink rate. It plunges earthward on a 20-degree slope, more than six times steeper than the approach of an airliner and faster than a falling skydiver. Shuttle landings are not for the faint of heart.

A ground-based microwave scanning beam landing system tells the shuttle exactly where it is during the final approach. With this support

↑ A night landing of the Space Shuttle *Columbia* at Kennedy Space Center. 1996.

→ Space Shuttle *Endeavour*'s parachute opens perfectly as it touches down on Runway 15 after an eight-day mission in space. 1998.

from the ground, the onboard computer is capable of auto-landing the ship, but as with the Apollo moon landings, the commander is entrusted with the option (always exercised) of manually controlling such critical touchdown maneuvers. The U.S. situation forms a sharp contrast to that of Russian spacecraft, which typically obey orders from mission control rather than allowing the pilot to have his own control; a cosmonaut takes over only in the event of an anomalous situation.

T-38 chase planes flown by other astronauts flank the incoming shuttle. The runway appears below in the rich green Florida scrub. Brilliant ground guide lights show the correct alignment and confirm the proper angle of approach. Two miles out, at 351 miles per hour, the commander pulls the nose up, exposing the belly of the orbiter to the wind and "flaring" the ship to a gentler touchdown speed and angle. Fourteen seconds from touchdown, the landing gear swing out—any earlier, and their aerodynamic drag would kill too much of the little lift the shuttle does enjoy. The rear wheels hit the runway at 220 miles per hour, and soon the nose of the ship dramatically drops down to land on its nose gear. A mortar in the tail fin fires a drag parachute, which blossoms to help slow the rolling orbiter and hold it straight as it barrels down the runway. Disengaging the chute at about 33 miles per hour prevents it from settling on the engine cluster, and the great craft brakes safely to a stop.

A space shuttle orbiter just in from space cannot be treated like an airliner. The chemicals and propellants used for the orbiter's engines, systems and maneuvering thrusters include poisonous toxins and explosive chemicals that may contaminate the area around the spacecraft. Hydrogen, hydrazine, monomethylhydrazine, nitrogen tetroxide or ammonia may be floating around the craft waiting to asphyxiate or burn someone, or even burst into flame. When the shuttle has landed, a recovery convoy of some 25 special vehicles and 150 trained people streams out from the waiting area alongside the runway. This team surrounds the ship with equipment designed to "safe" the orbiter. Before anyone takes off their protective gear, they check for dangerous leaks, and then hook up umbilical lines to off-load fuels and toxins. Purging air is pumped through the orbiter to clear and cool its payload bay and other cavities. The hookup of ground equipment allows onboard cooling systems to shut down. It is over an hour before the astronauts' hatch is opened and the space travelers step out, usually having already removed their orange pressure suits. They are hidden from view by a curtained ramp to their Crew Transfer Vehicle. The mission commander, sometimes accompanied by crew members, will usually walk around his vehicle for a final inspection review before it departs.

While the astronauts are driven off for their mission debriefing, the

The cockpit of the orbiter *Atlantis* showing the graphic flight indicator displays. 1998.

recovery crew hooks up a diesel tug connection to the shuttle nose wheel, and support personnel board the ship to replace the flight crew at the controls. The onboard personnel install switch guards to prevent accidents and gather data packages from mission experiments. The ground crew prepares the ship for safe land transport, fitting landing gear lock pins and checking over ship systems inside and out. A corresponding engineering team stationed in one of the firing rooms of the Launch Control Center monitors orbiter system data and transmits commands to prepare the shuttle for processing. Within four hours after landing, the recovery team begins to tow the spacecraft off the runway. A special tow-way carries the shuttle past egrets and alligators, and perhaps a manatee swimming the waterways. It is quiet again.

LANDING IN CALIFORNIA

The Edwards Air Force Base runways in California span the vast flats of a dry lake bed, giving test pilots a comfortable amount of maneuvering room and stable, predictable weather conditions. Edwards and the NASA Dryden Flight Research Center within its borders provided the landing

Immediately after landing, Mission Commander Tom Henricks inspects a tire on *Discovery*. 1995.

strips for the orbiter's development and early flights. However, once the test landings had proved out the system, NASA wanted the orbiter flying directly home to Florida, because it costs a fortune – three-quarters of a million dollars – to lift the orbiter up, put it on the back of one of NASA's two specially modified Shuttle Carrier Aircraft Boeing 747s, and fly it back to Kennedy Space Center. The ferrying process also costs five days' processing time, and exposes the precious $2 billion space shuttle orbiter to danger from hazardous weather or accidents along the way. Bad weather at the Cape can hold the orbiter in orbit for a while, but after power and oxygen supplies run too low, NASA must make the call and bring the ship in for an alternative landing.

White Sands Space Harbor in New Mexico serves as a second backup landing site and its Northrup Strip was used once to land STS-3, but the location's famous windblown white sand got into every crevice of the elaborate spaceplane and gave the KSC cleaning crew costly headaches. The White Sands landing site will only be used again in an emergency.

On its way back to Florida from a backup landing site, the Shuttle Carrier 747 carrying the orbiter has to stop repeatedly for fuel since the

The Space Shuttle orbiter *Atlantis* approaches the VAB where work will be performed in preparation for launch. 2003.

100-ton load is so heavy, but after a few days it arrives at the KSC Shuttle Landing Facility. A special 115-ton crane called the Mate/Demate Device waits on the landing strip's large parking apron, ready to lift the orbiter off the 747 and place it on the ground. One way or another, on its own or via carrier, the orbiter will always end up at KSC when its mission is completed, leaving the runway under tow.

PREPARING A SPACE SHUTTLE FOR LAUNCH

THE RETURN OF a space shuttle orbiter initiates a cycle of intense activity at Kennedy Space Center. Waiting technicians receive the spacecraft, render it safe after its dangerous trip, and then painstakingly prepare it for its next mission. Assembly crews then take over to combine the orbiter spaceplane with its rocket boosters, and finally roll the stack out to the pad. KSC teams have done this work over a hundred times, but every cycle involves elaborate tear-downs and a virtual rebuilding of the ship, so

complex is the orbiter's technology and construction. Every cycle builds toward preparing for a manned liftoff.

SHUTTLE ORBITER HANGARS

Like any other aircraft, the space shuttle orbiters have storage and maintenance hangars, but nothing at KSC has a name so simple as "space shuttle hangars." The set of three NASA garages for the space shuttle orbiters is accordingly known as the OPF, or Orbiter Processing Facility. Each hangar proper is an OPF "high bay."

Originally, when the space shuttle was envisioned as being more or less comparable to a high-performance military aircraft in its maintenance needs, these hangars were conceived as being much like conventional aircraft hangars. This was back in the days when the future space shuttle was often described as being able to land at regular airports, as Wernher von Braun's 1950s shuttle concept was designed to do.

When the final NASA shuttle design was settled and built, the real-world orbiter turned out to be fantastically complex, a very different machine than one that could land at airports. After every landing, an orbiter spaceplane must be torn apart and rebuilt to ensure that everything will work properly on the next mission – this takes more than an ordinary hangar.

In order to support this extensive process, the orbiters' maintenance hangars become packed to the rafters with elaborate work platforms, cranes and service equipment. When the orbiter is in the OPF, the spacecraft is literally buried under dense layers of gantries, to the point of being invisible.

Hangars 1 and 2, close to the VAB, are connected by a low bay. Hangar 3, across the street from the other hangars, was originally an orbiter modification and refurbishment facility, but it was eventually upgraded into another fully equipped standard maintenance hangar. The high bays of these sophisticated facilities stand almost 100 feet tall, and are 150 feet wide by 200 feet long. Adjacent low bays hold ranks of electronic, mechanical and electrical equipment, in addition to office space.

Within a few hours of its landing, an orbiter arrives by tow to a waiting hangar that has been readied for it. All the hangar's service platforms are withdrawn, crowding against the sides of the 29,000 square foot building to make room for the incoming shuttle. Once the shuttle is in, the doors are closed and the platforms move to enclose the vehicle. Bridge trucks use telescoping arms with rotating buckets to give technicians access to otherwise out-of-reach parts.

An orbiter just in from space is still a dangerous machine, even after the recovery crew has emptied its tanks. When it arrives in its hangar, the orbiter still carries traces of fuels, which are either toxic or flammable or both. The orbiter also contains dangerous pyrotechnic explosives, making the processing crew's priority the "safing" of the spacecraft. With special tools and protective gear, the technicians clean out any remaining traces of fuel and de-install the explosives. Only when fully safed is the orbiter truly "back from space." Now, as an inert vehicle, it can it be serviced and refurbished.

One of the first jobs the maintenance crew faces next is removal of any payloads still in the orbiter's cargo bay, after which they reconfigure the bay for the next payload. Far from being just a big open hold, the orbiter's cargo bay is fitted with special attachments and equipment for each mission. The maintenance crews must remove this gear and install the appropriate equipment for the next mission. Twin 30-ton overhead bridge cranes allow the technicians to move heavy components around the hangar with ease.

The orbiter *Atlantis* being towed along the blue turning lines into the Orbiter Processing Facility. 2001.

Work stands surround the orbiter *Endeavour*, allowing servicing and checkout of every part of the craft. 2003.

PROCESSING THE ORBITER

The job in the maintenance hangar is extensive and will involve the inspection, testing and refurbishment of systems throughout the spacecraft. The painstaking process involves a substantial team of people, and takes an average of almost seven weeks, more than two-thirds of the two and a half months it takes to process an orbiter between missions. Much of this ground processing work is carried out under contract by the United Space Alliance, a joint venture of Boeing and Lockheed.

Visitors to KSC can see from their tour buses that each orbiter hangar is wrapped in an octopus-like net of giant ductwork. This is the emergency exhaust system, which can quickly evacuate a hangar in the case of a highly toxic hypergolic propellant spill.

The orbiter service hangars double as garages where some modifications can be carried out, new technology installed, or design problems fixed. Major refits of an orbiter, such as the installation of a state-of-the-art "glass cockpit," require that the orbiter be flown home to Palmdale, California, location of the shuttle factory where these great ships were built.

In the Orbiter Processing Facility, tiles on the forward area of *Discovery* are checked and replaced. 2003.

Even when an orbiter has been shipped in from a refit at Palmdale, it is also completely inspected and fully verified at its hangar, in case something has been overlooked by the technicians and engineers in Palmdale, which has happened in the past. Propellant lines are purged, engines and thrusters are tested, and 24 major subsystems run through their paces, all the way down to checking the crew's cooking microwave and the toilet. Nothing is omitted in this exhaustive checkout of full system function and reliability. KSC has a special engine shop for the orbiters, and when the main engines are yanked out of an orbiter for servicing they are sent to this neighboring engine shop for work. Part of the testing process for the orbiter involves powering up its systems and running them through their paces. This part of the process is controlled from consoles in the Launch Control Center.

As *Columbia* so painfully demonstrated in 2003, the integrity of the orbiter hull and heat shielding system is critical. The system is made up not just of the famous tiles, but of over a dozen different kinds of materials. Reinforced carbon-carbon protects the nose tip and leading edges of the wings. Black tiles protect high-heat zones. White tiles protect medium heat zones. Every tile bears a catalogue number and has a unique shape. Tiles are permitted to endure a certain amount of wear before being replaced. Devising secure attachment for these tiles plagued the development of the shuttle until today's system was worked out: the tiles are glued on with rubbery glue painted on by hand. Although the tiles are rigid, their fixtures are flexible in order to accommodate the thermal expansion of the airframe. Gap-fillers between the tiles prevent them from "chattering" in the vibration of launch. Lower heat zones are covered in lighter weight materials to save weight for payload capacity. It comes as a surprise that like biplanes of World War I, the shuttle is covered in part with fabric. In some areas the lightweight heat shield system consists of a specially treated fireproof felt-like material that shields the aluminum hull of the shuttle hiding below all the coverings. The entire thermal protection system gets examined while the orbiter is in the hangar. It gets repaired, waterproofed or has elements replaced as needed.

When the shuttle is ready for its next mission, it is time for the payload to be loaded aboard. In the case of large, horizontal payloads like the European Space Agency's Spacelab mini-laboratory, the loading procedure takes place right in the orbiter hangar. Vertically-oriented payloads, like the Hubble Space Telescope, are maintained in canisters in a special vertical payload building, and are installed out at the launch pad.

When all is ready, the hangar chiefs sign off that the orbiter systems are prepared for space, and the orbiter is backed slowly out of the high bay. From here it travels the short trip to the towering VAB, which stands looming over the hangar zone. Within its depths the orbiter will be transformed from a landed aircraft into a readied spacecraft.

SHUTTLE STACKING

A space shuttle orbiter arrives in the VAB as Apollo spacecraft once did. In the same place that put together Apollo 11, the cranes and crews now assemble the composite space shuttle. The 250-ton Apollo-era bridge cranes have been replaced with new 325-ton cranes. Equipped with the industrial power of such equipment, the VAB crew can move gigantic space vehicle parts around like toys. Literally at the top of this whole operation is the VAB crane operator. This individual, working in a small cab in the attic of the gigantic building, is assisted during stacking by a whole team of spotters. Each spotter watches part of the procedure from a suitable vantage point, their positions scattered throughout the huge structure of the VAB. Carefully placed, each spotter is responsible for making sure no crane maneuver will accidentally knock a component into a wall, a girder or any other piece of equipment in this monstrous and complicated environment.

EXTERNAL TANK

A space shuttle orbiter has three main engines at its tail end, but you won't find any fuel tanks inside spacecraft for those engines. Getting up into space takes so much propellant that it would fill the entire orbiter to

An External Tank being transferred to a barge before being taken to Michoud Assembly Facility in New Orleans, where the foam that formerly prevented ice buildup on the tank's bipod fittings was replaced with heaters to reduce the chance of foam strikes to the orbiter. 2004.

overflowing. The shuttle's boost-to-orbit gas tank is a separate piece of the shuttle combination: the large mustard-colored container that the orbiter rides on, called the external tank, or ET, by NASA. The ET carries about half a million gallons of liquid nitrogen fuel and liquid oxygen to burn with it, all fed into the shuttle's three main engines through pipelines in the mounting brackets.

The external tank also serves as the structural backbone of a space shuttle stack. Everything is attached to the ET, and it must bear that weight both at rest and under the stresses of launch.

The external tank is basically a big empty can covered with spray-foam insulation, which gives it a rough texture. The insulation is similar to the kind available at hardware stores to insulate the walls of houses. It has a mustard yellow color to begin with, but it darkens with exposure to sunlight, aging toward a cinnamon hue. This is why external tanks look different colors in different photos—their shade depends on how old they are, how long they've been in the sun and what weather they've been exposed to. Only the first two ETs were painted white—since then paint has been left off, saving some 600 pounds of payload weight.

Like all rocket liquid propellant tanks, this main tank has two separate containers inside for the two parts of the propellant formula. The nose cone holds the liquid oxygen tank, while the main body of the tube holds the liquid hydrogen fuel. NASA packs most of the electronics for the external tank in the space between the two containers. The turtleneck of the ET is the strongest part of the tank, and to it are attached the hangers for the orbiter and for the two solid rocket boosters.

The ET has three attachment hookups for the orbiter, each roughly corresponding to the locations of the three sets of landing gear, with one at the nose and two at the rear. The nose attachment point, or bipod, serves as a bracket to hold the shuttle onto the stack. The rear two brackets also carry the vital pipelines for feeding propellants from the ET into the orbiter's main engine systems, for liftoff thrust. Other conduits built into these attachment brackets carry electrical hookups and pressurization lines, making them the most critical function points in the entire ET.

The external tanks arrive at KSC on transport barges that carry them from the same manufacturing plant in Michoud, Louisiana, that formerly made the Saturn V moon rocket first stages. Via canals the ETs float right into the heart of the space center, where they are offloaded onto a large trailer and trucked into the nearby VAB. The crane operators hoist them up to vertical and store them in checkout cells until the tanks are needed to build a space shuttle stack. The ETs are over 154 feet long, but the VAB is so vast that it can easily store a number of them.

A shuttle stack begins when twin solid rocket boosters are bolted onto a mobile launch platform. The "spine" of the stack, its external tank, is then hoisted into place, and mated to the boosters. The outrigger rockets are joined to the ET at only two places on each side, so they can be jettisoned easily when their work is done.

The final element is the orbiter itself. Picking up a $2-billion spacecraft with one hand is a job for a particularly trusted individual. To rate primary crane operator status in the VAB, you have to use the 325-ton crane to hold a felt-tip pen and mark an X on an egg on the floor 52 stories below you—without damaging the egg. If you can do that, you are ready to pick up an orbiter.

The crane operator locks onto a hoist bracket attached to both sides of the orbiter, and brings it off the floor. Its nose is raised gradually, until the entire spacecraft hangs vertical. Now it must be moved through the air and brought into contact with the external tank in exactly the right position. The crane operator has a view of what's happening below through the cab, the perspective the unique one of looking down at the nose of an orbiter hanging in space. The operation of mating the orbiter to the external tank usually takes five days.

An empty external tank falls to the ocean after separating from the space shuttle.

When the shuttle stack is checked out, the Apollo crawler comes in to shoulder the launch platform, as it has done so many times since the early days. Today the crawler sports a laser docking system that allows it to home in more precisely than ever on its exact parking position within the six pedestals that hold a mobile launch platform in the VAB. Crawler master Thurston Vickery keeps a close eye on the 3,000-ton behemoth crawler with its gargantuan engines, and expects precision from his drivers. "If someone came in half an inch off," he says, "I'd fire them."

The great VAB doors slide open, and the shuttle rolls out, beginning another journey aboard the crawler, down the crawlerway and toward the pads. Usually this takes place at night, and by dawn the spacecraft has arrived.

THE SHUTTLE PADS

A PRIMARY FUNCTION OF any launch pad complex is to fuel the rocket for launch. For the space shuttle, this means large quantities of liquid oxygen and liquid hydrogen. The launch pad plays an important role as a fuel depot, and its equipment pumps these propellants into the shuttle's external tank shortly before launch.

Each of the two component propellants is cryogenic, meaning that they are gases chilled so cold that they condense into liquids. Storing cryogenics requires special facilities that can maintain these fluids at super-cold temperatures — the liquid oxygen at −297°F, the liquid hydrogen much colder at −423°F. The liquid oxygen is stored in a ball-shaped tank that holds 900,000 gallons, the liquid hydrogen in a similar 850,000-gallon tank on the other side of the pad. To help keep the cryogenics cold, tanks are gigantic vacuum bottles, since vacuum blocks heat transfer better than any insulation material. The feed lines running from the tank up to the pad are similarly vacuum-jacketed, an exotic but necessary measure with these cryogenics. You may imagine the engineering challenge of building pipes that maintain their vacuum-tight seals perfectly, even when liquid comes gushing through at almost three hundred degrees below zero and the pipe metal suddenly shrinks with the cold. This is the kind of problem that rocketry takes in stride. Vacuum-jacketing on the liquid hydrogen transfer lines also prevents the local air from freezing solid around the pipes, a bizarre but real scenario.

When it is time to turn on the pumps and commit to fuelling the shuttle, the operation must be completed quickly. The external tank's foam coating is a poor insulator compared to the heavy vacuum-protected storage tanks and feed lines, so every minute the propellants sit in the ET,

← *Discovery* is towed form the Orbiter Processing Facility to the VAB in preparation for its 25th flight from Launch Pad 39B. 1998.

350,000 gallons of water are released at Launch Pad 39A during a test of new valves installed in the sound suppression water system that protects the shuttle against damage by acoustical energy. 2004.

they are warming up and boiling into gaseous form. Every gallon that boils off is lost, because it has to be vented before the tank pressure builds up, and it ruptures like a balloon warmed to bursting on a hot day.

Overpressurizing the main tanks is a danger with serious consequences. A ruptured ET on the shuttle pad, with an orbiter attached to the tank, would cause catastrophic destruction and probably explode the hydrogen in the main tank. To guard against the danger of overpressurization, special evacuator systems are in place to suck off the boiled-off gaseous propellant. The gaseous oxygen is harmless enough—it has been vented off in prelaunch white clouds ever since the days of the A4, all the way back to Peenemünde. Those white clouds floating around the early rockets and around the shuttle today are pressure-valve oxygen, the tank "blowing off steam."

The gaseous hydrogen has to be blown off as well, but it can't be done right at the pad since it would mix with the oxygen clouds, create a spontaneous ignition, and require NASA to rebuild Merritt Island after every launch. A smokestack-like tower within the pad complex vents and burns off this hydrogen, to ensure that flammable gas doesn't go floating back into the shuttle. This bright orange flame is visible like a torchlight before every shuttle launch. A fire so near the shuttle may seem a little alarming, but it is insurance against overpressurizing the ET and against stray gaseous hydrogen.

Boiled-off propellants are replaced by a constant inflow of reserve from the ready tanks. A lot more cryogenics must be on hand than will fill the tanks, since so much of it boils off. The shuttle pad liquid oxygen tank holds 900,000 gallons to fill a 143,000-gallon shuttle ET tank.

Twin pumps feed the liquid oxygen up into the nose cone of the external tank before launch. Each pump delivers 1,200 gallons of liquid oxygen per minute, the equivalent of a heavy fire engine's main deluge gun. Yet no pumps at all are required to move the liquid hydrogen. Forced into an extreme state, liquid hydrogen is a bizarre substance very different from familiar liquids like water. Hydrogen is the lightest element in the universe, so its liquid form is extremely light as well. A gallon of water weighs 8.3 pounds, but a gallon of liquid hydrogen weighs only nine ounces. Liquid hydrogen also has an extreme tendency to climb up the sides of a container through capillary action. It will quickly crawl right out of a test tube and run down the sides by itself. These properties mean that a small amount of hydrogen vaporizing in the storage tanks creates enough pressure to drive the lightweight liquid hydrogen all the way through the feed lines and up into the shuttle's external tank. The engineers have only to open a valve.

The transfer lines run up to the mobile launch platform, where they feed into the shuttle via the two tail service masts, instead of by way of an umbilical tower as with most other rockets. The tail service masts are two boxy gray pillars that hook into either side of the orbiter's aft flanks, into connections that are spacecraft versions of a car's gas flap and cap.

The service mast on the left feeds the hydrogen, the one on the right feeds the oxygen. The propellants run through the orbiter itself and out its black-tiled lower surface through the mounting brackets that hold the shuttle onto the giant external tank. The liquid hydrogen runs right into its interior tank at the bracket point, but the liquid oxygen has to run up a conduit line along the outside of the external tank to finally hit a black angle bend-cap and flow into the interior oxygen tank right in the middle of the ET's turtleneck, just below the nose cone. You can see this external liquid oxygen line and the prominent black bend-cap easily in many photos of the shuttle sitting on the launch pad.

Once it is full, the ET is the heaviest element of the space shuttle stack. Fully loaded, this bloated gas tank weighs in at 824 tons, or 1.6 million pounds. It takes plenty of power to get this off the ground, but the combination of five different rocket engines will deliver it.

The shuttle astronauts bunk in the same building that once housed the Apollo astronauts, the Manned Spacecraft Operations Building. They wave to the press on their way out of the building, all smiles, and board the "astrovan" for their trip to the pad. Technicians help them strap into their seats, a challenge since the shuttle is pointing straight up. The commander and pilot sit up front in the cockpit. Passengers sit behind them and below on the mid-deck, the shuttle's second level. One and all prepare for what is about to happen.

The overall shuttle design configuration is inherently dangerous in placing the delicate spacecraft so close to the blast of its own engines and the violent solid rocket boosters. To cushion the sound of the blast, NASA installed a major sound damping system on the shuttle pads, to absorb blast noise in a water cushion. Adding to the raw elemental energy at a launch of this magnitude, a whole collection of nozzles bursts into action 6.6 seconds before ignition. Together they deliver a flood worthy of Noah. This deluge appears in several locations: fountains along the ridgeline of the flame deflector in the flame trench underneath the shuttle, spouts in the launch platform holes through which the engines are about to fire, and six "rainbird" nozzles that flood the steel surface of the launch platform. A firefighter's large-size fire-hose delivers 250 gallons – an entire ton – of water per minute. Deluge guns atop heavy fire engines can shoot 1,000 gallons per minute. By comparison, peak flow at the shuttle pad hits a staggering rate of 900,000 gallons per minute. At the moment of ignition this water floods throughout the flame and blast noise impact zone, and, in addition to water bags stretched across the blast holes for the solid rocket boosters, takes the brunt of the energy, cutting it down considerably. The 290-foot-tall water tower that provides for the launch flood is easily visible standing next to the pad; the tank's entire 300,000 gallon capacity is exhausted in just 20 seconds. Its seven-foot-diameter pipes are big enough to ride a raft through.

The shuttle does not leave danger from acoustical energy behind when it rises off the pad. With all the flat, reflective surfaces and the spreading cone of energy from the rocket blast, the sound levels hitting the spacecraft actually peak when the shuttle is some 300 feet up. The shuttle will not be safely out of range of this threat until it is 1,000 feet high.

Nine seconds up and moving at 100 miles per hour as it clears the tower, the shuttle ponderously begins its "roll program," six seconds that dramatically spin the entire shuttle around to fly into orbit in proper orientation. By the end of the maneuver, fifteen seconds into the flight, the vehicle is rising at 188 miles per hour and has hit 2,173 feet in altitude. It gains speed rapidly, pressing the astronauts against their seats. To reduce the stress, at 28 seconds up, the main engines are throttled back by a full third. They are going 394 miles per hour and are a mile and a half high. The dynamic pressure of the buffeting slipstream builds and builds with the acceleration until, at about 51 seconds, it peaks. This is the point of greatest stress on the vehicle, just after it has broken the sound barrier. As it rises above the lower, thicker levels of the atmosphere, it surpasses the height of Mount Everest. Now the main engines can be throttled back up. The computer has the main engines roaring at 104% power at 59 seconds into the launch.

In line with the main engines, each solid rocket booster likewise varies its thrust, burning at rates preset by the shape of the hollow (star-shaped in cross-section) running down the core of the propellant inside the long white casing tube. The points of the star cause faster burning earlier on by exposing more of the propellant surface; as these points burn down there is less surface area and the thrust power declines.

A whole minute of full thrust follows, tripling the vehicle speed to 159,670 miles per hour. The astronauts are traveling at Mach 3.77, their entire vehicle now moving faster than a rifle bullet.

At 126 seconds, explosive bolts fire on the solid rocket boosters, slicing off their two connections to the external tank. Side-pointing thrusters fire powerfully to push the white cylinders away from the shuttle. The solid rocket boosters (SRBs) begin slowly to tumble. They will follow an arc, rising for a time before falling back to Earth from about 160,000 feet. They are only about two-fifths of the way up to space and will drop back without facing reentry. They're moving at just short of Mach 4, which is nothing compared to the Mach 22 that the shuttle will finally reach. Speeds like Mach 22 cause burn-up in the atmosphere. At less than Mach 4, the solid rocket boosters will tumble and quickly lose speed before their series of parachutes slows them and finally drops them into the ocean without damage. Two special ships equipped with divers and a submersible recover the empty rockets for return to the manufacturer, where they will be stripped down, refurbished, reloaded and returned to KSC for reuse.

The shuttle plows on with the power of its three main engines, burning through the liquid hydrogen in its external tank. Two more minutes pass, and the group on board passes the point of "negative return" as NASA puts it.

Four and a half more minutes pass with the main engines continuing to eat their way through all 383,066 gallons of the hydrogen fuel in the tank. Finally the tank is nearly dry, and the engines shut down at eight minutes, 33 seconds. The ship is moving at an incredible Mach 23.18, or 17,498 miles per hour, and is now 73.5 miles high. It has increased its velocity by over 14,000 miles per hour since it jettisoned the SRBs. This is what the high-energy fuel hydrogen will do.

This dramatic image, taken by Russian cosmonauts aboard their Soyuz transport vehicle, shows the Space Shuttle departing from the Mir Russian Space Station. The curvature of the earth can be seen in the lower right corner of the image. 1995.

In essence, the SRBs have lifted the shuttle up above the dense atmosphere to a thin-air environment, where the vehicle then pitches over to let the main engines build up orbital velocity.

In six seconds the engines are fully shut down and there is no more thrust. The astronauts have continued to gain altitude—another 4,000 feet—but they've lost a little speed to do it. They are now in orbit. The external tank has done its job; it is deadweight now, so it is cut loose like the SRBs at T+8:51. It drops off, deliberately tumbled so that reentry will tear it into small pieces that fall into remote areas of the Indian or Pacific Ocean. Once the ET is gone, the orbiter maneuvers using its onboard fuel tanks, which are very small.

LAUNCH ABORT

Astronauts have faced the danger of being trapped in a doomed craft ever since the day that a pilot first stepped into a space capsule at the Cape. Being strapped into a capsule surrounded by flammable fuels at the pad was the first arena of risk; either the space vehicle or the pad facilities could present deadly hazards in the case of fire or explosion. After launch, on the way into space at rocket velocities, an astronaut could no longer simply "hit the silks" and bail out with a parachute if something went wrong with the ship.

The first four test flights for the space shuttle included ejection seats for the two astronaut pilots, which would shoot them up and out through breakaway panels in the roof of the shuttle cockpit. After the proving flights were complete, however, the ejection seats were removed, since there would be no way for the passengers on the mid-deck below to eject in the case of trouble because they would be trapped by the deck above them. Jim Lovell of Apollo 13 knew all too well the real danger of mishaps aboard rockets, and he was among those who commented on the shuttle's lack of a launch escape system when the concept was under review. However, the requirement was cut because the shuttle program needed to save money, and it looked too expensive to design.

The *Challenger* space shuttle exploded shortly after launch in 1986. Highly unusual freezing cold launch conditions shrank the O-rings sealing the separate segments of the solid rocket boosters, allowing a tongue of flame, which subsequently burned into the external tank, to escape. The resulting hydrogen explosion was catastrophic, but a launch escape system comparable to that which had protected Apollo would have saved the crew, just as such a system had preserved a Mercury capsule from an Atlas rocket explosion after launch in 1961. The *Challenger* crew compartment survived the explosion intact, and three of the crew turned on their manual oxygen supplies after the blast, proof that they'd survived. Yet they were trapped and utterly powerless as the cabin fell for two and a half minutes before impacting into the ocean at over 200 miles per hour. This appalling experience was the price of ill-chosen economy. One might imagine that discovering the intact cabin on the ocean floor, along with evidence that the crew had survived the explosion, would have led to drastic redesign measures. But since the orbiter design was considered too thoroughly finalized without provisions for a detachable escape pod or cabin that could be pulled to safety like an Apollo capsule, the shuttle continues to fly without a launch escape system.

Wernher von Braun's original concept for the space shuttle put the space-

The crew of *Columbia* wave as they head to Launch Pad 39A for a simulated launch countdown. From left to right are: Ilan Ramon, Kalpana Chawla, Michael Anderson, David Brown, William McCool, Laurel Clark and Commander Rick Husband. 2002.

plane on top of its booster rocket, mirroring the placement of the Mercury, Gemini and Apollo capsules. This would allow the piloted spacecraft to escape away from the rocket behind it in case of a disaster, and would also, like the traditional capsule design, keep the delicate spacecraft far away from the violence of the main booster's rocket blast. Unfortunately the space shuttle's final design places the orbiter right next to the full violence of the blast. The piggyback configuration also places it in the path of danger from any particles detaching during launch from the external tank under the buffeting of the supersonic airstream. If the *Columbia* had sat atop its booster in the von Braun configuration, it would never have been exposed to the foam strike that proved fatal upon that orbiter's reentry. The foam would simply have broken harmlessly away.

The Apollo 1 disaster brought from NASA the vow and dedication that the redesigned Apollo Block II space capsule would be the safest spacecraft ever flown. In the shuttle era, the Apollo Block II capsule continues to rate, by far, as the safest spacecraft ever flown. Let us hope that the future will see NASA return to Apollo's higher standards, rather than continuing the compromises already made to the shuttle program.

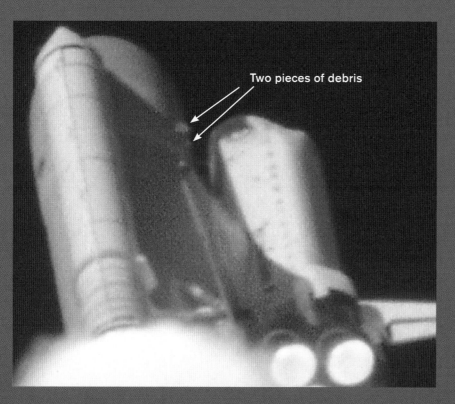

Two pieces of debris

← This low-resolution image shows foam insulation from the external tank striking Space Shuttle *Columbia* at the leading edge of the wing approximately 80 seconds into the launch. Damage to the thermal shield proved fatal on reentry. 2003.

↓ Workers attempt to reconstruct the bottom of *Columbia*. 2003.

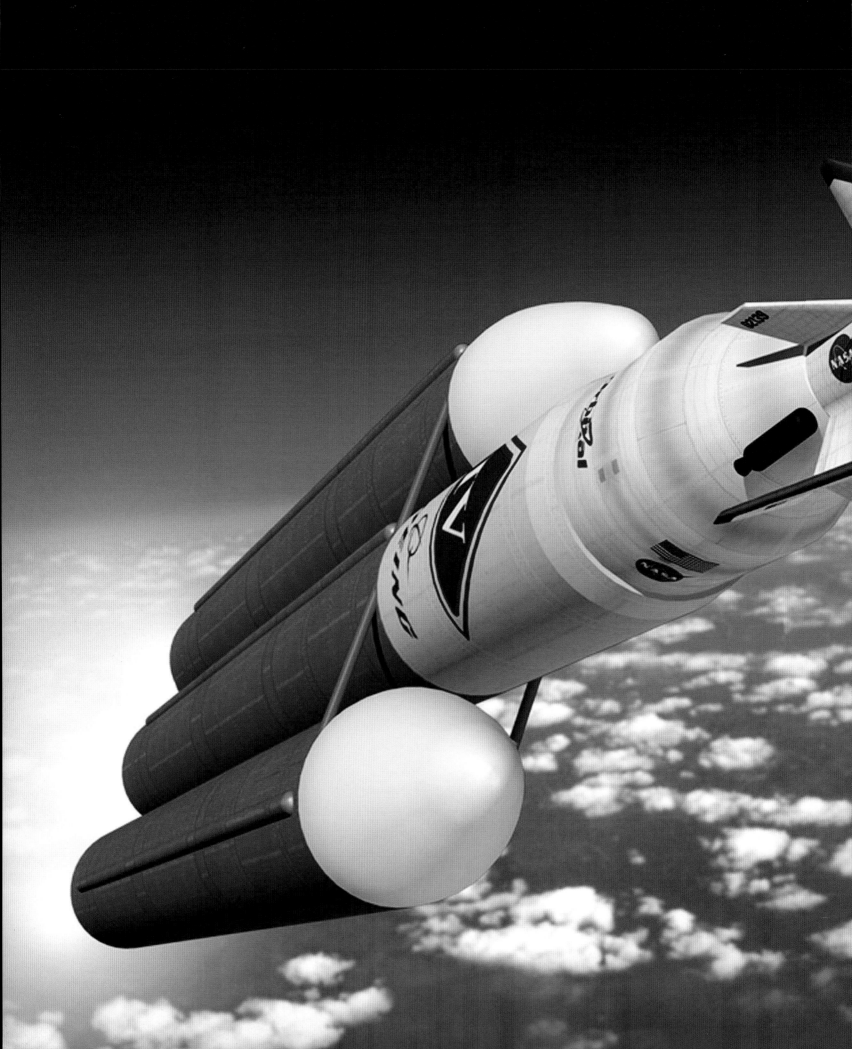

BEYOND THE SHUTTLE

A new era has dawned at the Cape with the announcement of a dramatic presidential initiative. The next generation of space exploration is finally taking shape, and the Cape is preparing for it.

PROBES AND SATELLITES

13

FOR EVERY ASTRONAUT launch from the Cape, there are many more unmanned launches. Unmanned rockets carry satellites for commercial communications, scientific research and military reconnaissance, as well as NASA space probes dispatched to scout asteroids, comets, Mars and the outer solar system. Known at the Cape as Expendable Launch Vehicles (ELVs), unmanned rockets range in size and capability, allowing payloads to be matched to appropriate carriers with no waste of excess rocket power. Several different contractors manufacture unmanned rockets, the leading U.S. corporations being Boeing and Lockheed Martin. The contractors operate their pads at the Cape under license from the air force, and carry out launches for commercial and military customers as well as NASA.

Most of today's unmanned launch vehicles can be traced back to origins in the 1960s or even the 1950s. It is so expensive and risky to develop an entirely new system that most successful rocket systems are based on previous successful systems. The lineages of the Cape's unmanned rockets illustrate what can come from good management of early investments.

← The Russian Mir Space Station as seen from the Space Shuttle *Atlantis* just after the shuttle completed undocking activity. 1996.

←← Artist rendering of the Joint Orbital Sciences and Northrop Grumman Space Taxi concept riding aboard an expendable launch vehicle.

FROM STOPGAP TO WORKHORSE: THOR BECOMES DELTA

THE THOR MEDIUM-RANGE missile was "too little, too late" in the face of superior Soviet rocket power in the late 1950s and early 1960s. Thors, designed as stopgaps until an intercontinental ballistic missile became available, were quickly rendered obsolete by the development of the more powerful Atlas. Within just a few years all the Thors were withdrawn from their foreign-soil sites around the Soviet Union, with the last British Thor coming home in August 1963.

The little stopgap missile seemed to have served its purpose, but that wasn't quite the end of it. The Thor was a good functioning system. Several "high-energy" upper stages were developed in the later 1950s, and Douglas engineers began adding these to the Thor to make two- and three-stage "combination" rockets. From these experiments, Thor was reborn as a probe and satellite launcher. Even the infamous Vanguard rocket's development was not wasted. The Thor-Able version topped the Thor with the old Vanguard second-third stage package (together called "Able"). This combination was powerful enough to launch the first U.S. probe toward the moon: Pioneer I got its ride on Thor-Able in 1958. Other Thor-Ables carried some of the Explorer scientific satellites and Tiros weather satellites, building an impressive record of accomplishment. The CIA took advantage of Thor power in a combination with an Agena-D upper stage, which had a restartable engine and offered precise repositioning once in space. Thor-Agena rockets slipped the top-secret Corona reconnaissance satellite, the first high-powered telescopic eye in the sky, into space, allowing the United States to peer into the military installations of potential enemy nations.

The best Thor combination rocket was the "Q-tip" Thor-Delta, which took the shortened name Delta in 1960. A Delta carried into space the world's first commercial satellite, AT&T's pioneering TV communications relay unit TELSTAR 1 in 1962, which introduced the TV phrase "live via satellite."

In 1963 the Douglas engineers gave the Delta combinations still more power by strapping on solid booster rockets to augment the liftoff thrust for heavier loads. With this set of options, the Delta became a flexible, customizable rocket system offering combinations of two or three stages with three, four, or nine strap-on solid rockets, depending on the power needed. This system became so effective and so reliable that Delta became the unsung workhorse of the space program, sending up many dozens of commercial satellites, the entire GPS satellite constellation, and many

→ This view from inside the Mobile Service Tower shows the Boeing Delta II second stage being redied to launch a deep impact space probe. 2005

**The Mars Explortion Rover 2
undergoes tests at KSC in
preparation for a mission to Mars.**

interplanetary space probes, including Mars missions such as the famous
Pathfinder which landed the Sojourner Rover on Mars in 1997, and the
Mars Exploration Rovers which landed in 2004.

Today the Delta series is still under development, with work carried on
by Boeing. The Delta II came online in 1989 for the first GPS satellite launch.
The Delta III first launched in 1998 had an unusual string of failures and was
dropped, but Delta IV is now ready to shoulder the heaviest cargoes.

At the same site where the first Thors debuted, Boeing currently oper-
ates Pads 17 A and B for military, NASA and commercial Delta customers.
In all, the Deltas continue to account for a great deal of activity on the
Space Coast, with Boeing's Delta operations at the Cape involving nearly
600 people. Boeing also sends Deltas into polar orbits from Space Launch
Complex 2E at Vandenberg in California.

Today the Delta II sets the standard for rocket reliability, boasting a
modern track record of 97 percent success. The Delta has become one of
the most commonly launched rockets of all time, with over 275 Deltas
launched since their origin in 1960. All this grown from a little failed stopgap called Thor, a very well-managed and superbly developed investment.

The Hubble Space Telescope 350 miles above the Earth can look into space without the distortions caused by the atmosphere.

ATLAS COMBINATIONS

WHILE LARGER ROCKETS were developed to carry astronaut spacecraft after the Mercury program used the Atlas to send the first Americans into orbit, the Atlas rocket continued its service for NASA probe missions. Just as with Thor, the addition of combination upper stages gave the Atlas a longer reach. This redirection gave the Atlas rocket a long and productive life beyond Gordon Cooper's last Mercury orbital flight. Atlas served NASA's second astronaut project, Gemini, by launching modified target Agena stages for the Gemini pilots to use in practicing space rendezvous maneuvers.

Atlas-Agena combinations also blazed the path for the Apollo lunar landings by sending a series of lunar probes to scout the moon. These included the early Ranger torpedo cameras, as well as the far more sophisticated Lunar Orbiter reconnaissance probes, which mapped the Moon with high-resolution photographic imagery. A Lunar Orbiter launched on an Atlas-Agena rocket identified the Tranquility Base Apollo landing site eventually targeted by Neil Armstrong's Apollo 11 lunar landing mission.

The third and final series of Apollo scout probes was Surveyor, but these large soft-lander probes were too heavy for the standard Atlas-Agena. A special hydrogen-burning high-energy stage called Centaur that provided enough extra power to launch the Surveyors was added. Launch Complex 36 was built to support the new Atlas-Centaur combination, which sent Surveyor 1 skyward on May 30, 1966. Surveyor 1's soft-landing touchdown on the moon's Ocean of Storms on June 2 produced 10,000 high-quality images of the lunar surface and proved that it was safe to make a landing on moondust. Other Surveyors followed, providing the program excellent familiarity with the lunar landscape before astronauts ever set foot there. Moon-walking astronauts would later catch up with the trailblazing Surveyor 3 probe and visit it during the landing of Apollo 12.

Atlas didn't stop at the moon. In combination with the Agena and later the Centaur upper stage, Atlas also launched NASA's interplanetary Mariner spacecraft, which made close flybys of Venus and Mars, Mariner 4 revealing the surprise of Mars' cratered surface for the first time.

At the dawn of the twenty-first century, the Atlas is still flying, several generations beyond that of the middle 1960s. In recent years Lockheed Martin has operated the Atlas family, fielding several versions of the Atlas-Centaur II and III from Pad 36B, while the air force has used Pad 36B for launches of military satellites on variations of Atlas IIs. Launch Complex 36 has kept busy with 10 to 12 missions per year and has flown an impressive tally of over 100 Atlas-Centaurs.

Lockheed Martin has further developed the Atlas into a fifth generation for heavy-lift power. The Atlas V that debuted in 2002 is the most powerful Atlas ever launched and has finally done away with the original balloon-tank design. Atlas V is such an extensive redesign of the Atlas system that it needed a whole new pad. In 1999 the air force handed Launch Complex 41 over to Lockheed Martin for conversion into a new complex dedicated to the Atlas V. The old Titan service and umbilical towers were leveled and a new assembly building constructed for this new generation of rockets built to shoulder 9.5 tons into space. The distinctive four white-tipped lightning towers surrounding the new Pad 41 are easily visible from several viewing sites in the Kennedy Space Center area.

Lockheed Martin completes nearly all the rocket's servicing inside its assembly building, moving its mobile launch platform out to the pad just 14 hours before launch. This is a substantial improvement over the weeks of traditional preparation required on the pad at Launch Complex 36. The short pad time allows for quick turnarounds and back-to-back launches to meet special scheduling needs. A new dedicated Atlas Spaceflight Operations Center building combines launch and mission control operations on-site at the Cape for these unmanned rockets.

Like Boeing's workhorse Delta, Lockheed Martin's Atlas V is a modular system with interchangeable parts that can be combined to meet the needs of specific payloads. One of the most surprising components of the Atlas is its variable-thrust engine, the RD-180, designed and built in Russia. The Russians have been building dependable engines so economically for so long that, on this side of the Cold War, they have finally broken into the American market. NASA believes in this international combination firmly enough to have hired the Atlas V to carry mankind's first probe to Pluto, the most distant planet in the solar system. The Pluto New Horizons launch, from Pad 41 on an Atlas V, lifted off in 2006.

FROM ASTRO-SOLDIERS TO THE OUTER PLANETS

ONE OF THE Cape's most elaborate launch complexes was originally built for launching astronaut-soldiers, but ended up fielding great missions of a very different type. The facilities for Pads 40 and 41 were constructed by the air force in the 1960s to launch the X-20 Dyna-Soar space bomber and the Manned Orbiting Laboratory military reconnaissance space stations. The Titan IIIC rocket devised to carry these heavy payloads needed its own new launch complex, and so the air force designed its own smaller version of the well-conceived Saturn V Apollo facilities, using a similar layout and similar architecture. The complex would even have its own junior VAB with shutter-doors. When the air force tried to find room for this elaborate complex, it found that all the ICBM pads had packed the Cape completely full. Undaunted, the air force went into the Banana River and dredged up artificial islands and a causeway to connect them. Through sheer determination, the Titan IIIC facilities emerged where there had been only shallows and sand spits before.

But the Dyna-Soar was cancelled in 1963, and then the Manned Orbiting Laboratory was cancelled in 1969 — the pair of programs rendered obsolete by the development of ICBMs and high-resolution reconnaissance satellites. Neither astro-soldier program ever saw a fully operational launch. The cancellations left the air force with an unneeded Titan IIIC heavy rocket and the elaborate 40/41 launch site that lay idle at the Cape. The Titan IIIC rocket was so overpowered that it could launch up to eight small satellites at a time. The military used it for various gang-loaded Defense Department satellite launches but never more than four per year. One pad lapsed into deactivation and it looked as if the whole complex was headed for mothballs until NASA gave it a second lease on life.

In the mid-1970s, the slow and over-budget development of the space shuttle left NASA without a heavy-duty launch system. Von Braun's Saturn IB rocket could have easily lifted such loads, but the Saturns were all gone. To get large interplanetary probes into space, NASA looked for the biggest lifter left in America and saw the Titan IIIC. Even the Titan IIIC was not powerful enough on its own for the ambitious mid-1970s mega-probes, but the rocket's design could be stretched a bit farther yet. In addition to the solid rocket power strapped onto its sides, NASA could add a high-energy Centaur upper stage. This steroidal combination would be mighty enough to hurl even a large probe all the way out of the solar system.

The swollen nose section resulting from the added Centaur stage looked too big for the Titan rocket underneath. But the combination turned out to be a viable solution—and more economical than a Saturn IB would have been. The "Titan IIIE" launched four of NASA's greatest exploration missions: two 1975 Viking missions to Mars and two 1977 Voyager missions to the outer planets. These probes returned the world's first look at Mars from ground level, along with astounding views of Jupiter (1979) and Saturn (1980). These were dazzling space exploration projects of the highest caliber. It seems fitting that after such grand missions to the great giant gas planets, Pad 41 has now dispatched a probe toward Pluto. Dredged out of nowhere and once meant for James Bond programs, the 40/41 complex has come to serve some of NASA's loftiest ambitions, and should see continuing use for years to come.

THE FLYING LAUNCH PAD

THE ORBITAL SCIENCES Corp. Stargazer is a totally different launch pad that uses the famous Skid Strip runway. This extraordinary "flying launch pad" is the only one of its kind, very much like something out of science fiction. The Cape is one of half a dozen launch locations around the world used by Orbital Sciences for launching small payloads into orbit using the Pegasus XL, a modern winged rocket. The Pegasus is carried into the sky for launch by a piloted, recoverable vehicle that takes advantage of aerodynamic lift to greatly reduce the thrust requirements of the initial boost phase. This unorthodox first stage—or flying launch pad—is the Stargazer, a Lockheed L-1011 airliner especially converted for the purpose.

Engineers have long known that you can give a rocket a flying start and save its fuel for longer journeys by dropping it from an aircraft, as long as the plane is big enough to carry the rocket. As so often happens in

↗ Workers install the second half of the fairing into place around the Solar Radiation and Climate Experiment satellite. The satellite was installed in the Pegasus XL launch vehicle. A specially modified L-1011 carried Pegasus aloft to 39,000 from Cape Canaveral Air Force Station. 2003.

→ A Pegasus launch vehicle being mated to an L-1011 prior to launch at Vandenberg Air force Base in California. 2004

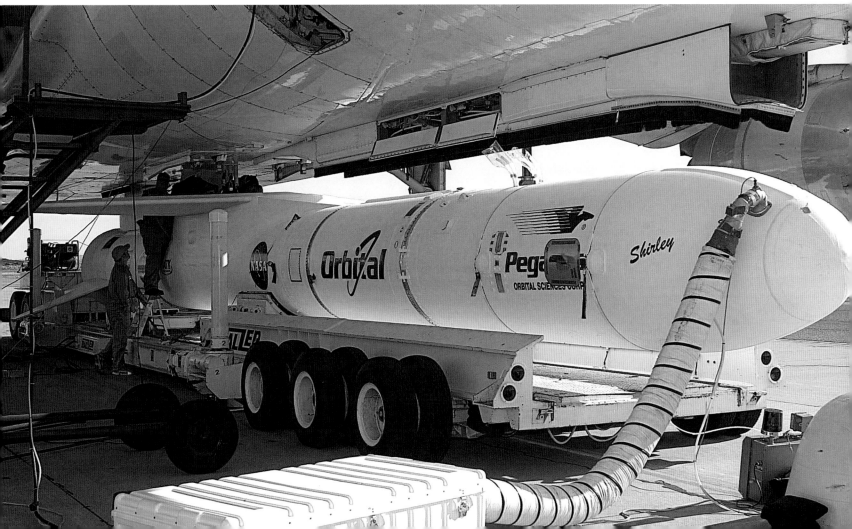

rocket science, the Germans tried it first. When the V-1 launch ramps were being bombed and captured by the Allies, the Luftwaffe sent up Heinkel bombers to carry the V-1s and drop them in flight. The head start offered by the bombers greatly extended the range of the V-1 and demonstrated a most intriguing technique. After the war, the U.S. tested many experimental rocket planes that also employed airdrop launch: Chuck Yeager's X-1 broke the sound barrier after being dropped from a B-29 in 1947; in the 1960s, Neil Armstrong and Scott Crossfield flew the incredible X-15 up to the edge of space after a similar B-52 ride to high altitude.

In the 1990s, Orbital Sciences found that a standard Lockheed passenger plane had the power to carry their 55-foot-long, 26-ton Pegasus rocket (nine tons heavier than an X-15). They modified an L-1011 airliner specially for the purpose, removing the passenger seats and adding a belly mount rack to hold their rocket tightly underneath the plane. This aircraft became the Stargazer. Thanks to the mobility of this unique flying launch pad, Orbital Sciences can stage launches from a variety of locations around the world, depending on the needs of a mission or a customer. Takeoff sites have included the Vandenberg Air Force Base in California, the Kwajalein atoll in the South Pacific, and the Canary Islands off Spain, but the Cape is one of the prime staging locations for Pegasus launches. The Stargazer takes off during a countdown monitored at the special mission control center for unmanned missions, located in Hangar AE. Once over the Atlantic Ocean 120 miles off Cape Canaveral, the Stargazer flies on a carefully plotted course, its progress constantly tracked with Global Positioning System monitors. For a successful launch, the launch aircraft has to hit a box-shaped 10 by 40-nautical mile launch zone at a calculated speed and 39,000 feet of altitude to release its rocket payload for a successful launch. The pilots must constantly fight the effects of weather interference to put the Stargazer through specific GPS milestones and into the launch zone on target, as if flying through invisible hoops to meet up with a swinging trapeze artist in the clouds with perfect timing. If any reading is too far off, the launch plane must cycle around for another pass.

When everything is lined up, the rocket is dropped. Letting go of a 26-ton payload is no casual event. In the words of copilot Donald Moor, "Launching a Pegasus is like unleashing a hypersonic freight train from underneath your aircraft." The entire Stargazer shudders with the bang of the four hydraulic hooks firing to release the rocket. Free-falling for five seconds, the Pegasus bursts into life with a roar you can hear on board the launch plane. A chase aircraft catches the ignition on video, as the Pegasus, now trailing flame, shoots ahead of the customized airliner, its 163,000-pound thrust engine quickly driving it through Mach 2 as it angles steeply upward toward its destination. The bright orange flame makes it easy to track as it heads off into the blue. From the cockpit you can even see the rocket drop its first stage at Mach 8, after about 70 seconds when it is almost 300,000 feet high. From there, the second stage engine can take the payload in the nose cone all the way into space, easily hitting 134 miles up, over 15 miles higher than the standard space shuttle orbiting altitude for missions below the space station. A third and final stage finalizes the orbital insertion. From airdrop to orbit is a trip of a little over ten minutes.

For small spacecraft and satellites that weigh in at less than half a ton, the Pegasus system is Earth's cheapest reliable ride to orbit: the cost for the entire launch is as low as $20 to 25 million. Together with the lift gained by the Pegasus rocket's 22-foot-wide wings, the air-launch system, with the "boost stage" being a recoverable piloted craft, makes the most of aerodynamic lift through the atmosphere. The system relies entirely on the brute force of the rocket engines only once the air has thinned out at upper altitudes. Taking advantage of the cheaper power of aerodynamic lift makes Pegasus a winner, and Orbital Sciences has reaped the rewards of their innovation by capturing contracts from customers around the world, launching more than 70 satellites on 35 missions since its debut on April 5, 1990. Today the Pegasus is the world's leading small launch vehicle—a particularly impressive achievement for the first privately developed space launch vehicle.

The Pegasus operations model the launch system long envisioned for an economical space shuttle. A piloted, plane-like recoverable booster stage has been seen as the ideal platform for a shuttle launch since way back in the 1960s. Such a winged first stage has never been built because it could not return its investment without a heavy schedule of constant launches...but if some future era brings greater need for frequent access to space, we may yet see some descendant of the Stargazer carrying astronaut and passenger craft up to airliner altitude for the dramatic beginning of their journey into space.

Wings are by no means rendered obsolete by the rocket age, and the Cape has a rich and ongoing history of winged launch vehicles to prove it. Winged rockets will continue to pass through these gateways to space.

GATEWAY TO THE FUTURE

14

Wᴴᴬᵀ ʟɪᴇѕ ʙᴇʏᴏɴᴅ the space shuttle for Kennedy Space Center? What spacecraft will carry the astronauts of the future to the high frontier? How will the Cape launch these new missions? For decades, space projects beyond the shuttle have been only the vaguest possibility. But a new era has dawned at the Cape with the announcement of a dramatic presidential initiative. New energies are at work, and NASA is designing new vessels for new deeds. The next generation of space exploration is finally taking shape, and the Cape is preparing for it.

← *White Knight One* circles the runway prior to landing at Mojave Airport Civilain Aerospace Test Center after carrying *SpaceShipOne* to its launch altitude. 2004

Artist's conception of NASA's new Crew Vehicle docked with Lander and Departure State in Earth orbit.

SPACE EXPLORATION INITIATIVE

ON JANUARY 14, 2004, President George W. Bush issued a remarkable document laying out his vision for U.S. space policy. This document, entitled A New Spirit of Discovery, directed NASA to phase out the space shuttle era and return to a mission of exploration. Specifying goals and timelines, and imposing executive restrictions on further use of the shuttle, the document made major decisions for NASA and provided, in the words of many observers, "the strongest presidential leadership for NASA since John F. Kennedy."

President Bush's initiative sets a bold new agenda for elevating NASA's goals beyond the shuttle's low Earth orbit, looking anew toward sending astronauts and robotic probes to explore the moon, Mars, and the solar system beyond. A Presidential Commission held hearings across the country and issued a report in June 2004, laying out a roadmap for accomplishing the President's orders. NASA began major reorganization and immediately set to the work of determining how best to accomplish the technical challenges laid out in the President's directives. New vehicles and new mission profiles needed to be designed. With NASA's workforce energized by their bold new mission, the work began quickly. The new space initiative has electrified NASA in a way the agency has not seen since Apollo. And, as ever, ground zero for the launch of this vision will be Kennedy Space Center, the great space gateway.

AFTER WORKING TO meet the recommendations of the *Columbia* Accident Investigation Board, NASA's first order of business was to return the space shuttles to operational status as soon as possible. Shuttle launches now have new safety measures in place such as heat shield repair kits that astronauts take into orbit on every flight. NASA is now on sharp lookout for anything like another *Columbia* launch damage incident, and has technology and procedures prepared to deal with such a development and to get the astronauts back safely.

Once back in action, the shuttle fleet operations were further strengthened by the one thing the program had never had before: a clear goal. President Bush directed space shuttle operations henceforth to focus on one objective: completing the construction of the International Space Station.

The space shuttle was originally conceived as a means of achieving a permanent space station in Earth orbit, an orbiting operations platform that would be a successor to the pioneering Apollo Skylab station. But in the early 1970s, NASA found too little support for such a complete program, and so it settled for the shuttle alone and hoped to get approval for a space station to "give the space shuttle someplace to go." President Ronald Reagan initiated serious studies on a space station design called Freedom, which would have been a space platform built out of small pieces that the shuttle could carry in its cargo bay, and that astronauts would assemble in space. This Space Station Freedom went through many designs and redesigns in an effort to find a configuration that would be large enough to be useful, but small enough to fit the budget available. The project languished without strong support and leadership.

SHUTTLE-MIR

IN THE 1990s NASA arranged a new partnership with Russia as an interim way to involve the shuttle with a space station, an arrangement developed from the seeds planted by the Apollo-Soyuz project. To give our astronauts practice in space docking and long-duration stays in space, we would send our space shuttles up to the Russian space station Mir, which had been launched in 1986 and expanded several times in subsequent years.

The Shuttle-Mir cooperative project brought NASA a great deal further experience in working with an international partner on space operations. Mir also brought more space adventures than we counted on. At one point the station caught fire, a very dangerous situation in such a confined environment. In another incident, an unmanned remote-controlled Progress cargo supply ship accidentally crashed into and depressurized a space station module that was intended to be a special United States laboratory. The collision greatly limited the amount of research the U.S. was able to conduct aboard Mir because the leak was never traced and plugged. Nonethless the program, which ended in May 1998, gave our astronauts space station experience in a time when we had no facility in space.

INTERNATIONAL SPACE STATION

IN THE EARLY 1990s NASA administrator Dan Goldin championed a new space station project using a formula that would finally bring success: internationalism. Goldin built support for a station that would be American-led but international in scope, with components built by partner space agencies in Europe and Japan.

After the breakup of the Soviet Union, Russia was invited to be a partner in the space station. To get Russian cooperation, the U.S. had to surrender control of the station and accept only half-time command in alternation with Russian command. But the station would get built, and the international angle would indeed end up locking the U.S. into a firm commitment, that this project would endure administration change, delays, and serious setbacks without being knocked off course.

The new space station was designed to serve as a micro-gravity laboratory, so that astronauts could carry out experiments in materials science, biology and physics.

Space station assembly flights began in 1998 and dominated launch operations at Kennedy Space Center thereafter, as the shuttle blasted into space again and again, every few months. The new station would be made up of numerous pieces small enough to fit in the shuttle's cargo bay, each of which would have to be joined to the construction via painstaking spacewalks often lasting many hours and setting records for duration work. A dwelling module, a power module, a crane and solar panels, piece by piece, were joined together at an altitude of 220 miles. The final design, with outstretched solar panels, would stretch 356 feet wide, almost

The International Space Station, photographed from the Space Shuttle *Endeavour*. 2001.

the length of a Saturn V moon rocket—longer than a football field. The completed facility was designed to weigh 475 tons, house a crew of up to seven people, and hold laboratories supplied by NASA, Russia, Europe and Japan.

One after another, these components have been blasted into the sky from Launch Pad 39A, adding to the increasingly large network of lattice-work and complicated engineering that constitutes the largest orbiting facility ever created. Today the space station is so large that you can easily spot it moving across the sky if you know where and when to look. It is one of the brightest objects in the night sky, brighter than most stars.

The International Space Station contains so many separate components that it will require several dozen space shuttle launches to complete construction. President George W. Bush directed NASA to work toward completing the station around the year 2010, which would require almost two dozen more launches from the shuttle fleet that was returned to operational status in 2004. As soon as the space station is completed, the shuttle fleet will be retired.

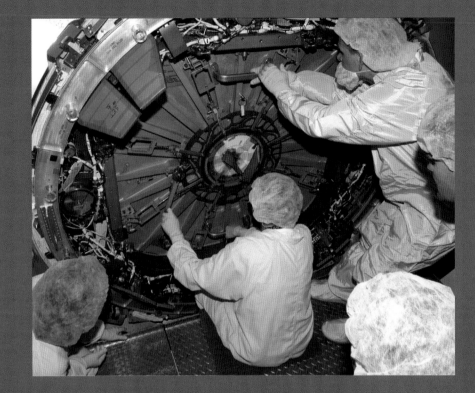

→ Technicians in the Space Station Processing Facility prepare to open the starboard hatch on the Node 2 module.

↓ In the Space Station Processing Facility, Executive Director of NASDA Koji Yamamoto points to other Space Station elements. Behind him is the Japanese Experiment Module.

PREPARING FOR NEW
LAUNCH VEHICLES

KENNEDY SPACE CENTER will be launching the shuttles for several years yet, but NASA is already working on developing exciting new vehicles that will form the next generation of astronaut spacecraft. It will be up to KSC to provide launch services for these vehicles, so just as during the design of the space shuttle, KSC is waiting to find out just what it is they will be launching. What facilities will be needed? Will the new vehicle require new pads, or can the existing equipment and facilities be adapted to the new purposes? A look ahead will give us some idea of how those questions may be answered at the Cape in the near future.

THE CREW EXPLORATION VEHICLE

President Bush's space initiative directed NASA to begin development of an entirely new space vehicle with the specific mission of supporting exploration: carrying astronauts beyond low Earth orbit, back out to the far frontiers around the moon, on the lunar surface, and then on to Mars. The new Crew Exploration Vehicle (CEV) resembles not a space shuttle but the more compact and economical Apollo space vehicle.

The twenty-first century Crew Exploration Vehicle will be similar to the Apollo design: a basic unit that can be customized and expanded with optional components to form a family of related vehicles capable of accomplishing various mission profiles. The CEV will face a wide variety of mission profiles, from retracing all the Apollo paths, to traveling all the way to Mars, to carrying astronauts to a Mars landing and possibly even beyond to the outer solar system.

NASA hopes that Crew Exploration Vehicle prototypes may be flying unmanned as early as 2008 or 2009, but it is quite possible that after the shuttles stand down around 2010 that the U.S. may have no astronaut spacecraft capability for several years, until man-rated CEVs are ready. During that interim, which will recall the six "down" years between Apollo-Soyuz in 1975 and the first Shuttle launch in 1981, NASA may make use of Russian Soyuz capsules to send astronauts up to the space station, a benefit of the cooperative space relationship the two nations have built in the years since their celestial Cold War competition.

NEW LAUNCH VEHICLES

When the Crew Exploration Vehicle is ready, the rocket that will carry it into space will determine where at the Cape these future missions may lift off. Both Boeing and Lockheed have proposed that a CEV could be

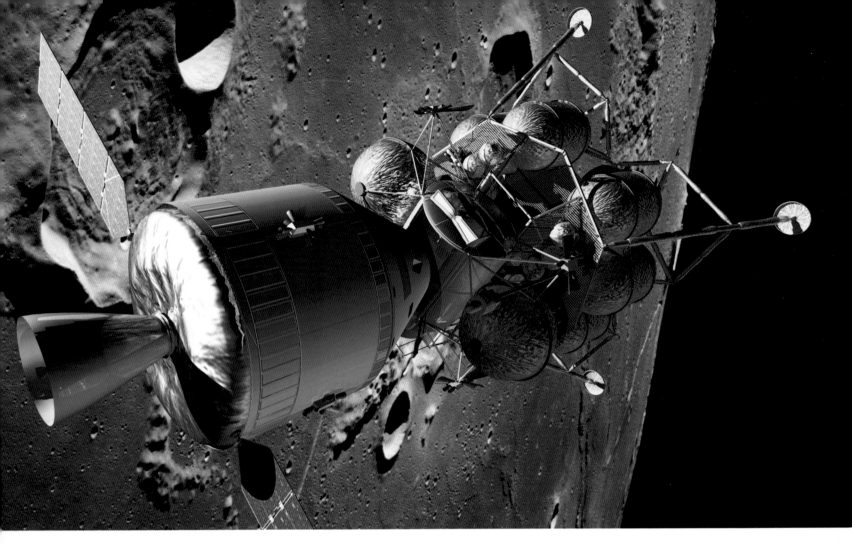

Artist's concept of possible new exploration vehicle for America's return to the moon.

launched on the heavy-duty variants of their existing launch vehicles, although these heavy-lift variants have not yet proven their reliability.

NASA may choose neither of these two existing rockets, opting for an entirely new booster that would essentially reincarnate the power of von Braun's Saturn IB and be, like that rocket, a vehicle designed especially for NASA's needs—and built from the beginning with the safety of astronauts in mind.

NEW CARGO-LIFTERS

In addition to launching the Crew Exploration Vehicle and the astronauts aboard it, NASA will need a cargo-lifter for the ambitious space exploration program. President Bush specifically directed NASA to separate crew and cargo flights as soon as possible, to avoid the problem of unnecessary expense that the space shuttle has engendered. Likely contenders for the cargo-lifter role are "shuttle-derived" vehicles such as the Shuttle-C prototype built by Boeing in the early 1990s. These vehicles use the building blocks of the space shuttle system and combine them in new ways that are more efficient and more powerful. Shuttle-C, for example, would use a standard space shuttle external tank and pair of solid rocket boosters, with the shuttle orbiter replaced by a large cargo tank fitted with spare shuttle main engines at the back. Without wings or life support systems to carry, this combination would have a capacity almost triple the payload capability of the shuttle hardware. A superior configuration has won favor by placing the cargo stage atop the vehicle, out of the way of damage by shed foam or ice in the slipstream, and by affixing the shuttle main engines at the base of the external tank. Other combination possibilities include using more than two solid rocket boosters, or new and more powerful booster stages. For the CEV launcher, NASA has examined the "stick" configuration, with a single SRB topped with an upper stage and the spacecraft.

All of these derived vehicle prospects have the advantage of using as their building blocks components that have been redesigned over the course of more than 100 shuttle launches; their engineering has already been proven and their weaknesses are already know. As we have seen time and time again, all advances in rocket power are dearly gained, and foolishly discarded. As mountain climbers say, "Never sacrifice altitude." In rocket science, this translates to "Never sacrifice hard-earned power capability."

Today we have the shuttle elements that are known and proven quantities. It is highly likely that NASA could make from them a good cargo vehicle to support our future astronauts' exploration missions, sending expedition equipment into orbit, to the moon, or to Mars. So even after the shuttle fleet stands down, its elements may continue in production. Shuttle engines now installed in active orbiters may be removed to power early shuttle-derived vehicles. Launch Complex 39A, adapted to support these vehicles as it was once adapted to support the original shuttle, may very well remain a highway to space well into the next decades, with exotic new hardware traveling out to the farthest frontiers from the same launch site that saw Neil Armstrong's historic liftoff on Apollo 11.

Meanwhile, NASA plans to launch robotic scouts to the moon by 2008, to prepare for astronaut exploration. A new series of manned lunar landings is scheduled to begin between 2015 and 2020, with the establishment of a base on the moon to prepare and practice for a longer expedition to Mars. After further robotic exploration of the "red planet," astronauts are to be dispatched to make a dramatic landing on that distant world. If the plan is upheld by future administrations, this leadership commitment will return NASA to endeavors of Apollo-era challenge and fascination.

Burt Rutan looks on as Paul G. Allen (center) congratulates Brian Binnie, test pilot, on completing the final flight for the Xprize challenge.

INNOVATIVE FUTURE LAUNCH POSSIBILITIES

As NASA looks ahead, proven orthodox technologies appear to be the most direct route to accomplishing dramatic exploration in the near future. Apollo achieved its success in a short time by keeping as much as possible within the engineering state of the art, rather than using risky and unnecessary experimental efforts. But while the main exploration efforts proceed upward from the Cape at full steam, gifted engineers will continue to develop innovative launch concepts that may yet hold keys to making space access easier, safer and less expensive for future-generation launch vehicles.

One of the intriguing possibilities is the rail launch system that has been explored by NASA's Marshall Space Flight Center in Huntsville, Alabama, under the direction of John London, who managed the Pathfinder Program that was designed to explore new kinds of space vehicles. A rail launch system could be built at KSC to send new types of launch vehicles skyward. "Air-breathing" vehicles get the oxygen to burn their fuel from the atmosphere, the way a jet engine does. This means that they do not need to carry heavy oxygen tanks, and lighter is always better in aerospace because it means there is less weight to lift. Air-breathers use wings to generate their lift during the time they are still in the atmosphere, like the Pegasus XL launch system does.

The rail system would use a magnetic levitation track to accelerate the vehicle magnetically. The rail launcher would pitch its vehicle off at 400 to 600 mph, a flying start toward space. At present there are no plans to build the rail launcher, but as NASA examines the best solutions to future launch needs, the rail launcher will figure high on the possibilities list.

SpaceShipOne

ON JUNE 21, 2004, the first privately built spacecraft took off from an airport in Mojave, California, and reached an altitude of 62 miles, 400 feet higher than the internationally agreed boundary that marks the beginning of space. World-famous aircraft designer Burt Rutan had turned his skills with lightweight composite fabrication to the problem of reaching space, and, with the funding of Microsoft billionaire Paul G. Allen, created a remarkable new form of spacecraft dubbed *SpaceShipOne*. Rutan also built a special turbojet carrier plane called *White Knight* to take the small three-seat *SpaceShipOne* up to about 46,000 feet for air-launch. *White Knight* would thus serve as a flying launch pad (or winged first stage) comparable to Orbital Sciences' airliner launch vehicle for the winged Pegasus XL rocket—for which Rutan himself designed the wings.

Mojave airport won a license that day from the Federal Aviation Administration as America's first inland spaceport, perhaps opening the way for many future such designations. Rutan's design included a new type of hybrid engine that uses safe, non-explosive fuels: a rubbery solid fuel slug is burned using nitrous oxide, and can be turned on and off, thus affording the easy storage of solid fuel with the control advantage of liquid fuel—and the disadvantages of neither. One result of this approach is that launch operations become immensely safer, and tens of thousands of people who came to watch the historic flight were invited to get within a few hundred feet of *White Knight* and *SpaceShipOne* at Mojave, a dramatic contrast to the miles of safety distance required between traditional launch vehicles and such audiences.

Rutan's achievement at Mojave demonstrated the possibility that suborbital space flights can be carried out on an extremely small budget—his entire program, including the mission control center, training systems, design, and construction of the carrier plane and spacecraft and all their new instrumentation, was accomplished for less than $30 million. Rutan predicts a coming era of space tourism with such suborbital hops to be followed by true orbital tourism and space hotel destinations, operated by private industry. If Rutan is right, in the future NASA may be able to contract with such operators for services to low Earth orbit, just as movie viewers saw Pan Am flying spaceplanes to an orbital hotel in Stanley Kubrick's *2001: A Space Odyssey*.

Will inland spaceports and space tourism diminish the role of the Cape in future space operations? If Rutan's predictions come true, spaceflight will gradually become more and more like commercial aviation, and will indeed launch from many locations besides Kennedy Space Center and the nearby pads of Cape Canaveral Air Station. However, the exploration of space at the high frontiers beyond Earth orbit will continue to require the most powerful vehicles in the world, and those vehicles will need the support facilities of the Cape. In coming decades, perhaps more and more ordinary citizens will come to see space for themselves, taking in suborbital and orbital tourism flights. Ideally we will see the private space industry supporting NASA's leadership of space exploration, the two coexisting in harmony just as seen in Kubrick's cinematic space vision.

Regardless of what happens with private space development, the high frontier of exploration beyond the earth will remain the province of explorer-astronauts for some time to come. The Cape will remain the prime departure point for missions of both robotic and astronautic exploration; KSC will continue to launch the NASA vision.

Wernher von Braun and Kurt Debus always built for the future. Kennedy Space Center and the Cape are rich in the results of that foresight. There is still room for Pad 39C, and the girders sticking out of the VAB roof are prepared to enlarge that building for wonders even greater than the Saturn V. If our ambitions ever rise to that height again, von Braun's foundation will be ready for them.

From this historic Space Coast location, we can look back on heroic milestones and look forward with confidence toward new legends in the making. The Cape has been America's Gateway to Space. Other space gateways are now opening, made possible by the pioneering work that took place here. While the Cape is no longer the only launch point into space, it remains, in all its towering glory, the gateway to the future.

→ Visitors are dwarfed by the Saturn V rocket at KSC's Apollo Saturn V Center which pays fitting tribute to our greatest accomplishment in space.

Kennedy Space Center Timeline

1940s

May 11, 1949
President Truman signs bill creating Air Force Joint Long-Range Proving Ground that would become Kennedy Space Center

1950s

July 24, 1950
First launch from the Cape, a modified V2 reaches an altitude of 10 miles

December 1, 1955
President Eisenhower designates Thor missile program "highest national priority"

October 4, 1957
Soviet Union launches *Sputnik 1*

December 6, 1957
Vanguard launched from the Cape

January 31, 1958
Explorer I, America's first satellite, is launched from the Cape

October 1, 1958
The National Aeronautics and Space Administration (NASA) begins operation with 8,000 employees and a $100 million annual budget

October 7, 1958
NASA initiates Project Mercury, America's first human space flight program

1960s

5 May, 1961
Allan Shepard makes first American space flight in a Redstone rocket

24 August 24, 1961
NASA acquires a further 80,000 acres at Cape Canaveral

February 20, 1962
John Glenn circles the earth in first manned Atlas rocket, *Friendship 7*

July, 1963
Construction begins on the Vehicle Assembly Building (VAB)

November 20, 1963
President Johnson renames Cape Canaveral facilities the John F. Kennedy Space Center after his fallen predecessor

March 23, 1965
Gus Grissom and John Young initiate the Gemini Program

April 6, 1965
Launch of first international communication satellite, *Intelsat I*

August, 1965
Construction of Crawlerway begins

January 27, 1967
Flash fire at Launch Complex 34 sweeps through the Apollo 1 capsule killing its three-person crew

November 9, 1967
Apollo 4, powered by Saturn V rocket, launches the Apollo program

October 11, 1968
Launch of Apollo 7, the program's first manned operation

July 20, 1969
Neil Armstrong becomes the first man to walk on the Moon

1970s | 1980s | 1990s | 2000s

May 24, 1972
President Nixon signs agreement for Apollo-Soyuz Test Project

July 23, 1972
Launch of LANDSAT I, first satellite to provide assessment of Earth's resources

December 19, 1972
Crew of Apollo 17 returns to earth, concluding the Apollo Program

May 14, 1973
Skylab I, NASA's first space station, is launched

July 17, 1975
Apollo-Soyouz docking

October 16, 1975
Launch of GOES I weather satellite

March 24, 1979
Columbia, the first Space Shuttle orbiter, arrives at KSC

April 12, 1981
First Space Shuttle launch

December 5, 1986
KSC's Payload Hazardous Servicing Facility is opened

January 28, 1986
Space Shuttle *Challenger* explodes seconds after lift-off

September 29, 1988
Space Shuttle flights at KSC resume following an extensive investigation into the *Challenger* disaster

April 24, 1990
Space Shuttle *Discovery* is launched to deploy the Hubble Space Telescope.

September 22, 1993
Discovery becomes the first Space Shuttle to land at KSC at night

June 23, 1994
Space Station Processing Facility opens as checkout point for elements of the International Space Station

June 23, 1997
First piece of international Space Station arrives at KSC

February 1, 2003
Space Shuttle *Columbia* explodes over Texas

June, 2003
First Mars's rover launched from Cape Canaveral to reach the planet in early 2004

January 14, 2004
President Bush unveils space program, including development of a new exploration vehicle and manned missions to the Moon

Directors of the Kennedy Space Center

1. Dr. Kurt H. Debus – 1961–1974

2. Lee R. Scherer – 1974–1979

3. Richard G. Smith – 1979–1986

4. Lt. Gen. Forrest S. McCartney – 1986–1991

5. Robert L. Crippen – 1992–1995

6. Jay F. Honeycutt – 1995–1997

7. Roy D. Bridges, Jr. – 1997–2003

8. James W. Kennedy – 2003

Recommended Reading and Online Resources

The following references offer additional information and online resources on the main subjects covered in this introductory book.

Books

Barbour, John. *Footprints on the Moon*. New York: Associated Press. 1969. Barbour provides an inspiring and exciting account of the missions leading up to and including Apollo 11. His story captures the wonder of the time and includes many details that have since dropped out of the standard histories.

Benson, Charles D., and William B. Faherty. *Gateway to the Moon: Building the Kennedy Space Center Launch Complex*. Gainesville: University Press of Florida. 2001. This is a paperback re-publication of Part I of *Moonport* (1978), the classic official NASA account of the building of KSC. Filled with details of engineering, politics and decision-making, the book offers a fully illustrated (black and white) insider's view of the construction of the Moonport.

———. *Moon Launch! A History of the Saturn-Apollo Launch Operations*. Gainesville: University Press of Florida. 2001. A re-publication of Part II of the classic *Moonport* (1978), this volume details the launching of the greatest rockets of all time – the Saturn-Apollo Moon rocket.

Bilstein, Roger E. *Stages to Saturn: A Technological History of the Apollo/Saturn Launch*. Gainesville: University Press of Florida. 2003. This is a paperback re-publication of the outstanding official NASA account (1980) of the Moon rocket's design and construction. While containing a wealth of technical detail, the book is also highly readable.

Bramlitt, E.R. *History of Canaveral District, 1950–1971*. So. Atlantic Dist. U.S. Army Corps of Engineers. 1971. This book provides a detailed and interesting account of the work that went into building the Cape facilities, from the point of view of the U.S. Army Corps of Engineers.

Brooks, Courtney G., James M. Grimwood and Llloyd S. Swenson Jr. *Chariots for Apollo: A History of Manned Lunar Spacecraft*. Washington, D.C.: NASA, 1979. This is an exhaustive official history of command and lunar module development and use through Apollo 11.

Chaikin, Andrew L. *A Man on the Moon: The Voyages of the Apollo Astronauts*. New York: Penguin. 1998. Chaikin's engaging book provides a composite memoir of the Apollo astronauts, built from many interviews. This is an excellent account of the astronauts' experiences, presented from their perspective.

Collins, Michael. *Carrying the Fire: An Astronaut's Journeys*. New York: Farrar, Straus and Giroux. 1974. This candid memoir by Apollo 11's command module pilot provides a good personal account filled with vivid descriptions.

Compton, W. David. *Where No Man Has Gone Before: A History of Apollo Lunar Exploration*. NASA: Washington, D.C. 1989. This is the official NASA history on reaching, and exploring, the Moon.

Godwin, Robert, ed. *The NASA Mission Reports*. Burlington, Ont.: Collector's Guide Publishing/Apogee Books (Space Series). 1999–2005 2004. The author reprints space mission press kits and includes post-flight astronaut debriefing transcripts and other related documents. These volumes also include a wealth of NASA diagrams and photographs, as well as CD-ROMS of images and motion pictures.

Harland, David M. *Exploring the Moon: The Apollo Expeditions*. Chichester, UK: Springer-Praxis Books. 1999. This book places the reader on the surface of the Moon alongside the astronauts. Harland follows the Apollo moonwalkers step by step and crater by crater. Also included are dozens of panoramic moonscape images assembled for the first time.

——. *How NASA Learned to Fly in Space: An Exciting Account of the Gemini Missions*. Burlington, Ont.: Collector's Guide Publishing/Apogee Books (Space Series). 2004. The author captures the drama of the Gemini adventure and clearly presents a wealth of technical detail.

——. *The Story of the Space Shuttle*. Chichester, UK: Springer-Praxis Books. 2004. This revised and updated definitive account of the Space Shuttle includes its scientific contributions, with mission-by-mission highlights and extensive reference tables. The book is thorough and highly readable.

—— and John E. Catchpole. *Creating the International Space Station*. Chichester, UK: Springer-Praxis Books. 2002. This is a meticulous and insightful account of the evolution of the design and construction of, and the rationale behind, the International Space Station.

Heppenheimer, T.A. *Countdown: A History of Spaceflight*. New York: Wiley. 1977. This solid, detailed overview of the history of space flights includes good coverage of the Soviet space program and a thoughtful analysis of the role of space exploration in politics and society.

——. *The Space Shuttle Decision: NASA's Search for a Reusable Space Vehicle*. Washington, D.C.: Smithsonian Institution Press. 2002. The book conveys the tortured political process that resulted in the design of the Space Shuttle and surveys the history of the vehicle's predecessors.

Launius, Roger D. *NASA: A History of the U.S. Civil Space Program*. Melbourne, Fla.: Krieger Publishing. 1994. This concise history of U.S. space exploration combines historical documents and narrative.

Murray, Charles, and Catherine Cox. *Apollo: The Race to the Moon*. New York: Simon & Schuster. 1989. Basing their work on a large number of interviews, Murray and Cox provide the best account of Apollo from the flight control and engineering point of view. The book focuses on behind-the-scenes political intrigue and backroom discussions.

Ordway, Frederick I. III, and Mitchell Sharpe. *The Rocket Team*. Cambridge, Mass.: MIT Press, 1982. The book provides a detailed account of Werner von Braun's rocket team: its formation in Germany, and its transfer to the United States for work at White Sands, Huntsville and the Cape. This sympathetic presentation benefits from the author's personal acquaintance with von Braun.

Schirra, Wally. *Schirra's Space*. Annapolis, Md.: Naval Institute Press. 1995. This is the memoir of the only one of the original Mercury Seven astronauts who commanded Mercury, Gemini and Apollo missions.

Wendt, Guenter, and Russell Still. *The Unbroken Chain*. Burlington, Ont.: Collector's Guide Publishing/Apogee Books (Space Series). 2001. Pad leader Guenter Wendt's memoir is a highly readable and enlightening personal account of astronaut launch missions and personalities, written from the pad technician's point of view.

West Reynolds, David. *Apollo: The Epic Journey to the Moon*. New York: Harcourt Books. 2002. This is a vivid history of the Apollo program, from its imaginative beginnings through the latest possibilities, with emphasis on the workings of the hardware. Included are specially commissioned cutaway illustrations of spacecraft and diagrams of Moon exploration routes.

Wolfe, Tom. *The Right Stuff*. New York: Bantam. 1983. Wolfe's vivid coverage of project Mercury is crackling with color and humor, even if not strictly accurate in every respect.

Online Resources

Kennedy Space Center offers a variety of accessible fact sheets covering the history and operations of its facilities. Official NASA press kits are also excellent sources of information, and most are available online. Selections are included among the resources below.

Apollo Archive
Kipp Teague's Apollo Archive serves as a comprehensive reference on the Apollo program and is especially strong for its collection of high-resolution images.
http://www.appolloarchive.com

Apollo, Gemini and Skylab press kits
http://www-lib.ksc.nasa.gov/lib/presskits.html

Apollo Lunar Surface Journal
Dr. Eric Jones has compiled the Internet's heavily researched Apollo Lunar Surface Journal, a fully annotated and illustrated record of the communications during the Moon explorations. This is an exceptional resource.
http://www.hq.nasa.gov/alsj/frame.html

Cleary, Mark. *The Cape: Military Space Operations 1971–1992*
The 6555th: Missile and Space Launches through 1970
In these two books, Air Force historian Mark Cleary provides a concise reference on military launches at the Cape, from the beginning through 1992, providing an important counterpoint to the better-known civilian space launches of NASA. The books can be hard to find, but their text is available online.
https://www.patrick.af.mil/heritage/Cape/Capefram.htm
https://www.patrick.af.mil/heritage/6555th/6555fram.htm

Countdown! NASA Space Shuttles and Facilities fact sheet
IS-2005-03-015-KSC
http://www-pao.ksc.nasa.gov/kscpao/nasafact/docs.htm

Crawler-Transporters fact sheet
FS-2006-01-001
http://www-pao.ksc.nasa.gov/kscpao/nasafact/docs.htm

Encyclopedia Astronautica
This is Mark Wade's excellent online space history encyclopedia. The "Cape Canaveral" entry is a good starting point for exploring articles on most of the Cape launch complexes and their rockets.
http://www.astronautix.com/sites/capveral.htm

From Landing to Launch (Orbiter Processing) fact sheet
IS-2005-06-018-KSC
http://www-pao.ksc.nasa.gov/kscpao/nasafact/docs.htm

Johnson Spaceflight Center
Johnson Spaceflight Center is home to "Houston Mission Control" and its online collection of digital images. This excellent resource is organized by program and mission.
http://images.jsc.nasa.gov/iams/html/pao/pao.htm

Landing the Space Shuttle Orbiter at KSC NASA fact sheet
FS-2000-05-30-KSC (2000)
http://www-pao.ksc.nasa.gov/kscpao/nasafact/landing.htm

Space Shuttle Transoceanic Abort Landing (TAL) Sites fact sheet
FS-2006-01-004
http://www-pao.ksc.nasa.gov/kscpao/nasafact/docs.htm

Spaceline
Cliff Lethbridge's Spaceline ebsite offers articles on the Cape pads, the rockets, and the earliest days of rocketry at the Cape.
http://www.spaceline.org/index.html

Wernher von Braun and the Early Years of Rocket Development
Wernher von Braun served as director of the NASA Marshall Space Flight Center. The center's heritage web pages offer articles on von Braun and the early years of rocket development.
http://history.msfc.nasa.gov/vonbraun/index.html

White Sands Missile Range fact sheets
The White Sands Missile Range Public Affairs Office provides many online charts, tables and articles, including coverage of the V-2, Tiny Tim and Bumper launches from the days when von Braun's rocket team was operating in New Mexico.
http://www.wsmr.army.mil/pao/FactSheets/fact.htm

Photo Credits

p. 107   NASA/JSC-S66-25782
p. 111   NASA/KSC-65-25875
p. 112-113   NASA/KSC-67PC-349
p. 114   NASA/KSC-72PC-176
p. 116   NASA Marshall Space Flight
        Center/MSFC-9131100
p. 117 ←   NASA/HQ-70-H-1075
p. 117 →   NASA Marshall Space Flight
        Center/MSFC-6200008
p. 118   NASA/GPN-2003-00056
p. 120   NASA/JSC-S-67-21294
p. 122   NASA/KSC-67P-0208
p. 123   NASA/KSC-64P-0145
p. 124   NASA Kennedy Space Center/KSC-04PD-2157
p. 125 ↑   NASA Kennedy space Center/KSC-05PD-1142
p. 125 ↓   NASA Kennedy Space Center/KSC-05PD-0179
p. 128   NASA/KSC-65C-0125
p. 129   NASA Kennedy Space Center/KSC-87P-0221
p. 130   NASA/KSC 98PC-1004
p. 131   NASA/MSFC-6870792
p. 133   NASA/MSFC6900558
p. 135   NASA/KSC-04PD-2685
p. 137   NASA/KSC-64C-5638
p. 138   NASA/JSC-AS11-40-5869
p. 142   NASA/KSC-69P-0623
p. 145   NASA Kennedy Space Center/KSC-69PC-0412
p. 146   NASA Marshall Space Flight
        Center/MSFC-6521237
p. 148   NASA HQ/GPN 2000-001483
p. 149   NASA/JSC-AS17-134-20384
p. 151   NASA/S70-17433
p. 152   NASA/HQ-SL3-114-1683
p. 154   NASA Marshall Space Flight
        Center/MSFC-8883912
p. 155   NASA/KSC-72PC-493
p. 156   NASA/KSC-73C-309
p. 157   © Bettmann/CORBIS
p. 159   NASA/JSC-SL3-115-1837
p. 160   NASA/MSFC-9401759
p. 162   NASA/JSC-AST-03-191
p. 163   NASA/JSC-AST-03-175
p. 164-165   NASA/KSC-02PD-1519
p. 166   NASA/DFRC-ECN-8607
p. 169 ←   © Bettmann/CORBIS

p. 169 →   NASA/MSFC-9132000
p. 171   © Dean Conger/CORBIS
p. 173   NASA Marshall Space Flight
        Center/MSFC-9142273
p. 176   NASA/KSC-05PD-0518
p. 179   NASA Kennedy Space Center/KSC-04PD-0937
p. 181 ↑   NASA/KSC-96EC-1336
p. 181 ↓   NASA/KSC-98PC-255
p. 182   NASA/KSC-99PP-0412
p. 183   NASA/KSC-95EC-1053
p. 184   NASA Kennedy Space Center/KSC-03PD-3201
p. 185   NASA Kennedy Space Center/KSC-04PD-0454
p. 186   NASA/KSC-01PADIG-139
p. 187   NASA Kennedy Space Center/KSC-03PD-2516
p. 188   NASA/KSC-03PD-3213
p. 190   NASA/KSC-04PD-2160
p. 191   NASA/Roger Ressmeyer/CORBIS
p. 192   NASA/KSC-98PC-1062
p. 194   NASA/KSC-04PD-1063
p. 198   NASA/JSC-STS071-S-075
p. 200   NASA/KSC-02PD-1987
p. 201 ↑   NASA/KSC-03PD-0243
p. 201 ↓   NASA/KSC-03PD-0733
p. 202-203   Getty Images Entertainment
p. 204   NASA/ STS79-E-5327
p. 207   NASA/KSC 04PD-2663
p. 208   NASA/KSC 03PD-0789
p. 209   NASA/Hubblesite.org
p. 212 ↑   NASA/KSC-03PD-0162
p. 212 ↓   NASA/KSC-04PD-2324
p. 216   Doug Benc/Getty Images
p. 218   NASA/John Frassanito and Associates
p. 221   NASA
p. 222 ↑   NASA/KSC-04PD-0637
p. 222 ↓   NASA/KSC-03PD-1954
p. 224   NASA/KSC S99-04195
p. 226   © Mojave Aerospace Ventures LLC,
        photograph by Scaled Composites.
        SpaceShipOne is a Paul G. Allen Project.
p. 229   NASA Kennedy Space Center
p. 232-233   NASA S-65-61887
p. 234-235   NASA/Reuters/CORBIS

Index

Abernathy, Silas, 141

Aero Spacelines, 31

A4 rockets, 50-55

Aldrin, Buzz, 110, 144, 148

Allen, Paul G., 227

Apollo program, 32, 33, 97, 113, 115, 123-24, 150

    Apollo I, 119-120, 200

    Apollo 7, 120, 121, 161

    Apollo 8, 23, 161

    Apollo 11, 130, 131, 134, 141-48, 209

    Apollo 12, 150, 210

    Apollo 13, 144, 150

    Apollo 14, 150

    Apollo 15, 150

    Apollo 16, 150

    Apollo 17, 150

    Apollo 20, 154

    Apollo Block II, 200

Apollo-Soyuz Test Project (ASTP), 160-63

Armstrong, Jan, 145

Armstrong, Neil, 110, 144, 145, 148, 169, 214

Assembly No. 3 (A3) rocket, 49

astronauts, 98, 108, 109, 119, 148, 199

    International Space Station, 220

    Original Seven, 79, 81, 82, 83, 86-87, 89, 94-95

    Skylab, 154, 158

    space shuttle, 167, 172, 175, 181, 196, 220

Atlas rockets, 39, 40, 89, 209

    Atlas V, 26, 210, 211

Atlas-Agena rocket, 106, 209

Atlas-Centaur rocket, 210

Atlas Spaceflight Operations Center, 210

barge canals, 33, 126, 191

Boeing, 26, 27, 177, 187, 205, 208, 223

Borman, Frank, 108, 109, 110

Bucyrus-Erie, 126

Bumper rockets, 37, 39, 62, 63, 64-65, 147

Bush, George W., 218, 221, 224

California Institute of Technology, 61, 77

Cape Canaveral, 21, 23, 27, 62-63

Cape Canaveral Air Force Station, 23, 25

cargo-lifters, 224

Carpenter, Scott, 95

Central Intelligence Agency (CIA), 206

Cernan, Gene, 110, 150

Chaffee, Roger, 119

Challenger, 199

CIA (Central Intelligence Agency), 206

Clinton, Bill, 158

Collins, Michael, 144, 148

Columbia, 157, 188, 200, 219

computers, onboard, 98, 110, 181

Conroy, Jack, 31

control centers, 39, 40, 58, 64, 71, 89, 99

Convair, 89, 106

Cooper, Gordon, 87, 95, 106

Corona reconnaissance satellite, 206

cosmonauts, 81, 87, 160,181

crane(s), 33, 103, 127, 189

      operators, 33, 35, 189, 191

crawler-transporters, 23, 35, 124, 126, 132, 134-36, 193

crawlerway, 23, 35, 135

Crew Exploration Vehicle (CEV), 223-24

Crossfield, Scott, 214

cryogenic liquefied gases, 36, 50, 60, 193, 195

Dannenberg, Konrad, 50, 53, 54-55

Debus, Kurt, 39, 69, 71, 77, 124, 126, 127, 228

Delta Mariner, 33

Delta rockets, 26, 206, 208

      Delta II, 208

      Delta III, 208

      Delta IV, 33, 208

Disney, Walt, 168

docking, 110

      ring, 161

Dornberger, Walter, 49, 52, 55

Douglas Aircraft, 31, 33, 62, 74, 206

Dryden Flight Research Center, 178, 182

Dyna-Soar rocket plane, 170-71, 211

Edwards Air Force Base, 182-83

Eisenhower, Dwight, 74, 76

ejection seats, 108-09, 172, 199

engines, rocket, 53, 60, 89, 98, 116-17, 143, 211

escape towers, 86, 108, 127, 145

European Space Agency, 189

Expendable Launch Vehicles (ELVs), 205

Explorer I, 77

Faget, Max, 94, 172

flame deflectors, 39, 103, 117, 136, 196

flame trench, 39, 50, 136, 196

Friendship 7, 90-91

fuel. See propellants

fuel cells, 98

fuel slug, 227

Gagarin, Yuri, 81-82

gantries, 37, 39, 50, 58, 64, 72, 99-100, 101, 103, 119, 136, 186

Gemini program, 97-101, 103, 104, 106-10, 209

      Gemini 5, 106

      Gemini 6, 106-07, 108

      Gemini 7, 107-08

      Gemini-Titan II, 30, 103

General Electric, 58

Glenn, John, 25, 81, 87, 89-91, 94, 103

Global Positioning System (GPS) satellites, 26, 206, 208

Goddard, Robert, 45, 61

Goldin, Dan, 220

GPS (Global Positioning System) satellites, 26, 206, 208

Gray, Norris C., 64

Grissom, Gus, 81, 87, 103, 119

Grumman, 143

Guggenheim Foundation, 61

Ham, Linda, 157

hangars. See also Vehicle Assembly Building

      "alphabet," 33

      space shuttle, 185-86, 187

hatches, 119, 120

heat shields, 91, 94, 157

      shuttle, 188-89

Hubble Space Telescope, 189

hurricanes, 129

hydrogen peroxide, 50

hypergolic fuel, 103, 108

ignition, 39, 53, 54, 60, 74, 87, 90, 98, 129, 136-37, 146-47

inter-continental ballistic missiles (ICBMs), 25-26, 39, 40, 89, 98

International Space Station (ISS), 157, 161, 219, 220-21

Jet Propulsion Laboratories, 61, 62, 77

Johnson, Lyndon, 23, 143

Joint Long Range Proving Ground, 43, 62-63

Jupiter C, 77

Kennedy, John F., 23, 97, 124, 143, 147
Kennedy Space Center (KSC)
    location, 23, 62-63
    naming of, 23
    size, 23, 29
kerosene, 82, 108, 137, 147
Khrushchev, Nikita, 87
Korean War, 71

Lark winged missile, 39
Launch Control Center (LCC), 39, 40, 139, 140, 174, 175
launch pad(s), 25-27, 36, 49, 58, 77, 99, 101
    crew of, 81, 82, 84, 101, 104, 106
    flying, 212, 214, 227
    Pad 3, 63, 64
    Pad 5, 25, 69, 71, 72
    Pad 6, 69, 71, 72
    Pad 14, 25, 107
    Pad 17B, 74
    Pad 18A, 76
    Pad 19, 107
    Pad 34, 117, 119, 120, 120, 121, 154
    Pad 36B, 210
    Pad 37, 119, 154
    Pad 39A, 23, 136-37, 221, 225
    Pad 39B, 23, 154, 155
    Pad 40, 35, 211, 212
    Pad 41, 35, 210, 211, 212
Leonov, Alexei, 162
Ley, Willy, 45, 168
liftoff, 36-37, 39, 54, 60, 64-65, 74, 76, 87, 91, 103, 109, 146-47
liquid hydrogen, 191, 193, 195, 197
liquid oxygen (LOX), 50, 60, 82, 137, 190, 193
Lockheed, 27, 187
Lockheed Martin, 26, 205, 210, 211, 223
London, John, 226
Lovell, Jim, 108, 109, 199
lunar lander, 127, 143, 148
lunar probes, 209, 210
lunar roving vehicle, 150

Mach numbers, 91
Malina, Frank, 61
Manned Orbiting Laboratory, 211
Manned Spacecraft Operations Building, 25, 196
Manned Spaceflight Center, 172
Mariner spacecraft, 210
Marion Power Shovel, 126
Mars missions, 208, 210, 212, 223, 225
Marshall Space Flight Center, 115, 226
Martin Company, 32, 98, 99, 100, 101, 103, 104, 107, 108
Matador winged missile, 39
Mate/Demate Device, 184
McDonnell Aircraft, 81, 104, 107
Meilerwagen, 58
Mercury program, 33, 39, 40, 81-84, 86-91, 94-95, 97
    Mercury-Atlas, 89-91, 94-95
Merritt Island, 23
    National Wildlife Refuge, 27
Mir space station, 219-20
missiles, military, 25-26, 39, 40, 73, 89, 98
Mission Control, 40, 140
Mississippi Test Facility, 33
Mobile Launch Platform, 130, 195
moon landing missions, 97, 98, 106, 110, 225. See also Apollo program
Moor, Donald, 214

NASA, 23, 86, 97, 101, 121, 167
    and armed forces, 27, 103, 104, 170-71
    rescue capability, 157-58, 199, 200, 219
    and U.S. space policy (2004), 218-19, 223
Navaho winged missile program, 171
Naval Research Laboratory, 76
Nebel, Rudolf, 143
New Spirit of Discovery, A, 218
Nixon, Richard, 150, 167
North American Aviation, 119, 120, 144, 150, 174
NOVA rocket, 116
nuclear missiles, 73. See also inter-continental ballistic missiles

Oberth, Hermann, 48, 143
O'Keefe, Sean, 158
orbital navigation, 109-10
Orbital Sciences, 212, 214, 215
orbiter, space shuttle, 167, 168, 172, 175, 177, 185-86
Orbiter Processing Facility (OPF), 185-86, 187
O-rings, 199

Pathfinder, 208
Peenemünde rocket center, 45, 49, 50
    Test Stand VII, 50, 52, 53
Pegasus XL rocket, 212, 214, 215, 227
Pershing missile, 26
Pioneer I, 206
Pluto New Horizons mission, 211
Port Canaveral, 26, 33
Pregnant Guppy, 31-32
propellants, 36, 60, 82, 98, 108, 144, 181, 191, 193-94, 195
purging gas, 36

rail launch system, 226-27
rail transport system, 30, 35, 126
Range Safety office, 27
RD-180 engine, 211
Reagan, Ronald, 158, 219
Redstone rockets, 25, 26, 39, 40, 71-72, 79, 86, 87
re-entry, 91, 94, 179
rendezvous in space, 106-107, 108, 109-10
    American-Russian, 159-63
rescue rockets, 156-57
robotic scouts, 225
rocket planes, 169-70, 214
Rocketdyne, 89, 116, 143
rockets, space. See also names of individual rockets
    assembly of, 30, 33
    launch, 36-37, 39
    monitoring and controlling, 40
    preparation for launch, 36-37, 39
    shipping of components, 30-33
    transport of, 35
Rockets (Levy), 45
rollout, 132, 134

R-7 ballistic missile, 74
Rutan, Burt, 227, 228

safety measures, 27, 36
    shuttle, 167, 172, 186, 188, 199-200, 219
    spacecraft, 81, 84, 86, 103, 104, 106, 108, 120, 156-58, 161, 227
satellites, 26, 74, 76, 77, 162, 205, 206, 208, 210, 211, 215
Saturn rockets, 30, 117
    Saturn I, 32, 116, 119
    Saturn IB, 33, 35, 116, 119, 121, 154-55, 161, 212
    Saturn V, 23, 33, 35, 36, 37, 39, 116, 123, 126, 127, 129, 130, 131, 132, 136, 140, 144, 146-47, 153, 154, 167
Schirra, Wally, 87, 95, 108, 109, 110, 120, 121
Schmitt, Harrison, 150
Shepard, Alan, 25, 81, 82, 83, 84, 86, 87, 103
Shuttle Carrier Aircraft Boeing 747s, 183-84
Shuttle Landing Facility (SLF), 178, 179, 181
Skid Strip, 30, 32, 212
Skylab, 32, 153, 154-58, 163
Slayton, Deke, 87
solar eclipse, artificial, 163
sonic boom, 137, 178
Soyuz spacecraft, 160, 161, 162, 163, 223
space capsules, 81, 98. See also Apollo program
space industry, private, 227, 228
space probes, 205, 208, 209, 210, 211, 212
Space Race, 77, 88, 90, 104, 108, 116, 141
space shuttle, 30, 120, 167
    backup landing sites, 182-84
    design, 168-72, 199-200
    external tank (ET), 189-91, 193, 194, 195, 197, 198
    future of, 218, 219, 221
    image, 175
    launch, 196-98
    launch abort, 199-200
    launch pad complex, 174, 174, 175, 193
    launch preparation, 184-91, 193-96
    and Mir space station, 219-20
    orbiter recovery, 177, 178-79, 181-82
    payloads, 186, 189
    solid rocket boosters (SRB),197, 198

stacking of, 189

Space Station Freedom, 219

Space Station Processing Facility, 23

space tourism, 228

Space Transportation System. See space shuttle

SpaceShipOne, 227-28

spacewalks, 106, 110, 220

Sputnik, 74, 76

Stafford, Tom, 107, 108, 110

Stargazer flying launch pad, 212, 214

Stennis Space Center, 33

STS-3, 183

Super Guppy, 32

Super Loki rockets, 27

Surveyor probes, 210

TELSTAR 1, 206

Tessman, Bernard, 71

Thiel, Walter, 52

Thiokol, 30

Thor missile, 73-74, 206

Thor-Able rocket, 206

Thor-Agena rocket, 206

Thor-Delta rocket, 206

Tiros weather satellite, 206

Titan rockets, 25, 98-100

    Titan II, 30, 116

    Titan III, 35

    Titan IIIC, 170, 211, 212

    Titan IIIE, 212

Titov, Gherman, 87-88

Toftoy, Holger, 61, 62

towers, 86, 108, 127, 145

    umbilical, 36-37, 106, 127, 130-31, 174

Truman, Harry, 43, 63

T-38 chase planes, 181

Turner, Harold, 60

umbilical towers, 36, 106, 127, 130-31, 174

United Space Alliance, 187

U.S. Air Force, 25, 26, 27, 35, 73, 74, 210, 211

U.S. Army, 26, 61, 71, 72

U.S. Army Corps of Engineers, 40, 63, 119

Usedom Island, 49-50

Van Allen, James, 77

Vanguard rocket, 76, 77, 206

Vehicle Assembly Building (VAB), 23, 33, 35, 127, 129, 132, 174, 189

Vertical Integration Building (VIB), 35

VfR club, 48

Vickery, Thurston, 134, 136, 193

Viking missions, 212

von Braun, Wernher, 31, 58, 69, 82, 88, 115, 228

    and Redstone rocket, 71, 76, 77

    and Saturn rocket, 117, 129, 141, 143, 147, 153

    and shuttle design, 168, 199-200

    and V-2 missile, 45, 46, 48-49, 50, 52

von Karman, Theodore, 61

Vostok I, 82

Voyager missions, 212

V-2, American, 30, 35, 36, 57-60

V-2, German, 45, 46, 50-55

Wac Corporal rocket, 61-62

Wainwright, Loudon, 147

water deluge systems, 37, 103, 196

Webb, James, 86

Wendt, Guenter, 84, 104, 108, 119, 120-21, 134, 144

White, Ed, 106, 119

White Knight, 227

white rooms, 39, 101, 104

White Sands Missile Range, 57, 58, 60-61, 62

White Sands Space Harbor, 183

World War II rockets, 45-46, 50-55

X-1 rocket plane, 214

X-15 rocket plane, 169-70

Yeager, Chuck, 91, 214

zero gravity experiments, 154

Zoike, Helmut, 53-54

zone of isolation, 94

Photo Captions

Cover The STS-34 Space Shuttle *Atlantis* lifts off from Launch Pad 39-B marking the beginning of a five-day mission in space. *Atlantis* carries a crew of five and the spacecraft Galileo, to be deployed on a six-year trip to Jupiter. 1989.

2-3 A spectacular lightning storm over the VAB scrubbed the launch of the Space Shuttle Endeavour, which was scheduled to take a replacement crew to the International Space Station. 2002.

4-5 Space Shuttle Discovery sits on the Mobile Launcher Platform and the Crawler-Transporter as it approaches the Rotating and Fixed Service Structures on Launch Pad 39B. 2005.

6 The spent Apollo 6 interstage falls away, as photographed by a camera on the second stage that would soon detach and parachute into the Atlantic. 1968.

9 This satellite image shows Merrit Island and the Cape jutting into the Atlantic Ocean as they appear from space. 1989.

12-13 NASA and Manned Spacecraft Center officials join the flight controllers in celebrating the successful conclusion of the Apollo 11 mission. From left foreground, Dr. Maxime Faget, George Trimble, Dr. Christopher Kraft Jr., Julian Scheer (in back), George Low, Dr. Robert Gilruth and Charles Mathews. 1969.

232-233 With the aircraft carrier *USS Wasp* alongside, Navy frogmen assist the recovery of the Gemini 6 spacecraft just minutes after splashdown. 1965.

234-235 Space Shuttle *Discovery* rides piggyback atop a specially modified Boeing 747 on a flight from California to Kennedy Space Center. 2005.